Future of Work-G7 Nations

Trends, Insights and Automation Effects

Future of Work-G7 Nations

Trends, Insights and Automation Effects

Dr. Mythili Kolluru **Dr. Mavis Gutu**

walnutpublication.com
INDIA • UK • USA

Hardback ISBN: 979-8-89171-258-4

eBook ISBN: 979-8-89171-260-7

First Published in June 2026

Published by Walnut Publication

(an imprint of Vyusta Platforms Private Limited)

www.walnutpublication.com

India

Unit# 909, 9th Floor, Wave Silver Tower, Sector-18, Noida - 201301

UK

71-75 Shelton Street, Covent Garden, London, WC2H 9JQ, UK

Distributed by

ZopioTail

About the Authors

Dr. Mythili Kolluru

Dr. Mythili Kolluru is an award-winning educator, strategic management scholar, and global learning architect with 23+ years of experience across the Middle East, Asia, and North America. She currently works as an Assistant Professor at the College of Banking and Financial Studies in Muscat, Oman. She has designed, delivered, and evaluated MBA, undergraduate, and executive programs, leading curriculum innovation, quality assurance, and accreditation-aligned teaching and assessment. In addition to being a certified Strategic Planning Professional, Dr. Kolluru is also a high-impact corporate trainer, consistently rated 4.9/5. A prolific researcher and writer, she has published 30+ peer-reviewed papers in globally indexed journals (including Scopus/Q1–Q2 outlets), authored specialized books, and created copyrighted frameworks such as the Strategic Thinking Wheel and the Personalized Strategic Career Decision-Making Grid. As a columnist with 60+ Oman Observer articles, she translates complex ideas on strategy, leadership, productivity, international affairs, and the future of work into accessible insights for policymakers, industry, and the public.

She can be reached at https://www.linkedin.com/in/drmythilikolluru/

Dr. Mavis Gutu

Dr. Mavis Gutu is a lecturer, and a Global Online Programmes Specialist in the Business School at Edinburgh Napier University, Edinburgh, UK. She is a Senior Fellow (SFHEA) of the Higher Education Academy, and an Associate (ACGAI) of the Chartered Governance and Accountancy Institute. Her research focuses on key current trends such as future of work, international business and marketing, and foreign market entry strategies. She has published in ABS4* and ABS3* journals. She has presented at both local and international levels.

Acknowledgements

The journey of writing this has been a humbling experience for both of us. Although our names are chiselled on the cover, many others have contributed with their generosity, wisdom, and encouragement.

To begin, we are both thankful to each other. In today's world, finding a trustworthy, hardworking, and committed researcher is a rare privilege. Working together has been a high-performing team experience, where we complemented one another in brainstorming, writing, delegation, and providing constructive and candid feedback, and in overseeing many stages, including editing, proofreading, and publishing.

We acknowledge the contribution of Walnut Publication, which have supported the book from the very beginning. Special thanks to Sharadha for patiently answering our relentless barrage of questions. The editor's feedback on refining the book is appreciated.

In this exciting journey from "Idea to Print", our respective families have played a significant role in discussions, idea generation, and the provision of critical comments on the book. Dr. Mythili's offerings of gratitude, a special thanks to my husband, Shyam Kumar, my die-hard fan, and my most unflinching critic. His honest feedback has sharpened my ideas, and his encouragement has been my steady anchor. During our long online marathon sessions reviewing the manuscript, he not only offered insightful suggestions but also lightened the atmosphere with cups of tea and lively conversation with both of us. The greatest joy of my life is my son, Vivek. From stocks to stars, our endless discussions fuel my curiosity and joy. As a voracious reader himself, he often shares unique perspectives that subtly find their way into my writing. My heartfelt thanks also go to my sister, Shanti, and my dear friends, Shobhana and Dr. Roslin, who have been unwavering pillars of encouragement and motivation throughout this journey. My sincere thanks to the College of Banking and Financial Studies for fostering an ecosystem that supports innovation in teaching, research, and community engagement. Finally, I offer a humble Namaskaram to my late, beloved parents, my earthly guides, and to my Lord, Arunachala, whose grace continues to enlighten my path.

Dr. Mavis is grateful to her loving husband, David, whose support and encouragement are relentless. His external discussions were a good distraction, allowing her to step away from it all and re-engage with a fresh perspective. To my son Ryan and my daughter Rayna, thank you for bearing with my endless talks on the book and related topics. Your sense of humour is encouraging. You made the long journey seem short.

To my eldest son, Randy, and my eldest daughter, Samantha, your current work experiences on this topic were interesting and set the stage for what the future of work could look like. My thanks goes to The Business School of Edinburgh Napier University, for cultivating a cultural research environment that promotes and supports innovative knowledge creation, sharing, and community engagement. Most importantly, Dr. Mavis is thankful to God Almighty, the Giver, Sustainer, and Director of all good things.

Preface

I am very judicious when committing to work with anyone on any project. My work and life experiences have taught me well that there is more to a person than their qualifications. An individual's personality traits and attitude are essential, as they determine compatibility and collaboration among team members. I met Mavis over LinkedIn, and we associated and communicated for a long time.

During these research meetups, we discussed what work looked like 20 years ago and how much it has changed since then. Moreover, we joked about how work will be for our children and our grandchildren. When we were discussing this topic, I got the "Eureka Moment." I jumped up with excitement. Why don't we write a book on the future of work? At first, it was overwhelming. Again, after much discussion, we dropped the idea and decided to write a paper first, then write our book later. A lull followed this for quite some time. Eventually, we came back to the same point; it felt like a maze. After much debate, we decided to write a book instead of a paper.

Do you want to know why you are working now and how you might work in the future? If the answer to this question is yes, then this book is primarily written for you. You might be a lawyer, a data scientist, a schoolteacher, a nurse, a doctor, or an ordinary person; the book offers takeaways for readers of all kinds. "WORK" has phenomenally changed over time and is changing as you read this book. So, hold on to this book tightly as a bumpy journey of work transformations across the G7 nations unfolds the changing face of work. The book will change your perceptions of work and reveal trends, developments, and the effects of automation, offering a useful insight into the future of work.

Table of Contents

Chapter 1

Introduction

Future of Work Context

Organisations facing post-COVID-19 disruptions are rapidly adopting technology and automation to manage market competition, operational uncertainty, and cost pressures, thereby fundamentally redefining the future workplace. This transformation is characterised by interconnected global trends: economic volatility, industry restructuring, significant demographic shifts, and the convergence of traditional and digital business models. Moving forward, the workplace will experience economic fluctuations, including demand volatility, industry reorientation, substantial demographic shifts, and the merging of previously distinct industries. Organisations will experience increased demand for digital skills, leading to new work models and requiring revisions to employment structures and novel international work permit arrangements, such as digital nomad visa provisions. These dynamic changes to the organisation and work landscape will have significant environmental implications, requiring collaborative dialogue among governments, the private sector, and organisations to address climate change and implement mitigation strategies. The workplace will witness a fundamental redesign of its job architecture, job roles, and career pathways, making them flatter, more dispersed, and increasingly influenced by industry disruptions such as automation, digital transformation, and remote working. Growth through innovation will impact organisational leadership models, which could require agile leadership approaches and leaders who can leverage advanced technologies to make sustainable decisions aligned with future-of-work models and that prioritise the well-being of all ecosystem stakeholders.

The future of work is characterised by fluidity and adaptive capacity, responding to major disruptions that fundamentally reshape organisational and individual work practices. Changes in future-of-work models will have significant effects on workforce composition and distribution. These changes will require constant reskilling and upskilling initiatives, new models of leadership, hybrid work arrangements, and digital transformation. The shift will alter the definition of what constitutes industry, introducing new meanings to the concept of "organisation". Variations in work patterns, such as remote working, represents one of the most substantial workplace transformations since the COVID-19 pandemic. Recent research demonstrates substantial disagreement between managers and employees regarding hybrid work implementation and emerging patterns (Bloom et al, 2023; Franklin, 2021). Employee perceptions of work have shifted substantially. Contemporary employees increasingly seek recognition, valuation, and meaningful engagement, moving beyond instrumental views of work as purely transactional and viewing work as central to personal identity and wellbeing (Franklin, 2021; Harter & Wigert, 2025; Kantar, 2022).

Diversity and Inclusion

Globalisation has created unprecedented geographic mobility in the workforce, necessitating that organisations actively value diversity and implement inclusive practices to attract and retain global talent. Diversity and organisation inclusion play increasingly prominent roles in workplace strategy; moreover, sustainability has become essential in reshaping physical workspaces. As hybrid and remote working models become increasingly accepted, technology emerges as the critical enabler in developing comprehensive workplace strategies. Uncertainty and dynamic workplaces have prompted employees to demand choice and control over when and where they work. Employees now increasingly seek control over their work arrangements to improve work–life balance. Digitalisation and cloud adoption will continue to enable the development of flexible and adaptive work models. Wellbeing will replace productivity metrics as the leading factor driving workforce engagement and organisational strategy (Trenerry et al, 2021; Attaran, Attaran, & Kirkland, 2019). The top strategic focus areas are employee safety, wellbeing, and enabling flexibility through employee-preferred working arrangements. Emerging organisational strategies in the future will increasingly be focused on employee wellbeing, lifestyle preferences, and the design of physical and virtual work environments.

In 2024, the gig economy market size was worth $556,7 billion and is projected to triple by 2032 to $1,847 billion (Charlton, 2024), as the gig economy continues to shape the work environment (Clark Ashar, 2025). Even

LinkedIn's predictions indicated a 150% increase in remote job postings by 2025. However, projections for 2026 show a decrease in remote job postings from the peak of the pandemic, but the level will be almost the same as in 2019. The jobs will be increasingly concentrated in experienced and high-level roles. The growth is particularly prominent in knowledge sectors such as software, consulting, design, and marketing. The rapid expansion of artificial intelligence (AI) technologies is evidenced by substantial adoption rates among freelancers, with approximately 60% of independent workers adopting AI-enabled tools and platforms. Market projections indicate that the creator economy is positioned to reach substantial market value, potentially exceeding $400 billion by 2027 (Forbes, 2025; Goldman Sachs, 2025).

The evolving work landscape presents complex challenges across multiple dimensions. Cybersecurity risks are escalating as organisations undertake digital transformations. Variations in labour, health, and safety regulations across countries create remuneration complexity, whilst geographic pay differentials present novel human resources management challenges (Bielefeld et al, 2025; Lazarova et al, 2023). All stakeholders, governments, organisations, and individuals, must collaboratively develop frameworks that align workforce development with organisational sustainability and individual wellbeing. Contemporary organisations require flexible, adaptive work environments that facilitate collaboration and innovation. Rigid, standardised approaches are insufficient; organisations must develop differentiated talent management strategies aligned with diverse workforce preferences and organisational objectives.

Implications of Artificial Intelligence for the Future of Work

Artificial Intelligence (AI) is often described as a disruptor of the future of work and carries implications for all stakeholders. It is critical for employees to gain a better understanding of AI's role in the future of work, which could facilitate growth, boost confidence, and inspire people to perform better together.

AI implementation must be conceptualised not merely as a technological investment but as a comprehensive people-centred strategy. An effective and efficient deployment of AI in the workplace is one that envisions AI working alongside employees to aid efficiency and effectiveness in carrying out their roles, rather than the objective of replacing human workers. Employee anxiety regarding potential job displacement stems partly from uncertainty. Organisations bear responsibility for transparent communication regarding AI implementation strategy and its implications for workforce composition and role evolution.

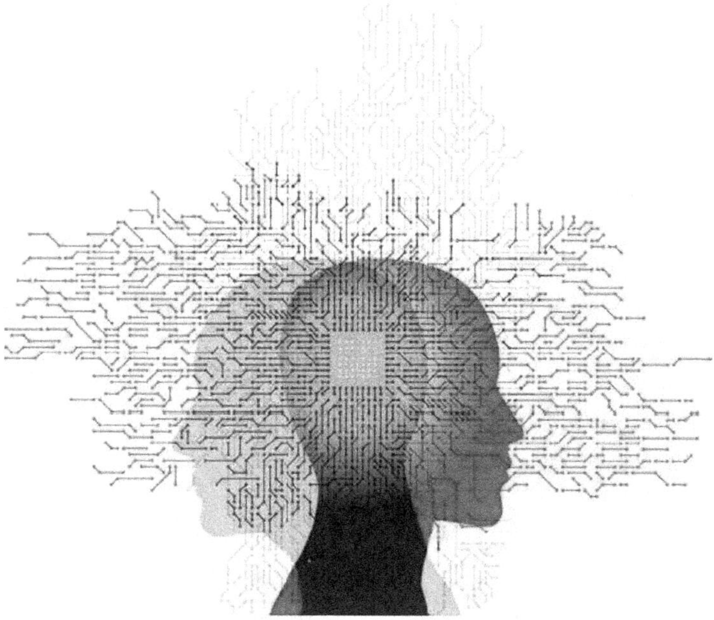

Figure 1.1: AI image (Source: Getty Images)

Organisations can foster an AI collaborative culture that could be instilled to enable organisations, enabling humans and AI to work together and dispelling employees' anxiety about the potential loss of their jobs. Analysis from the PricewaterhouseCoopers (PWC, 2025) Global Jobs Barometer found that artificial intelligence adoption enhances worker value in highly automatable sectors. Over a billion job advertisements were analysed across six continents to examine AI's global impact on jobs, wages, productivity, and skills. Key findings include (PWC, 2025): a threefold increase in revenue per worker in AI-exposed sectors; substantially increased AI adoption across previously unexposed industries such as mining and agriculture; accelerated skills change in AI-influenced roles; and a 25 % wage premium for workers with advanced AI capabilities compared to peers in equivalent roles without such exposure. These findings suggest that AI implementation, whilst necessitating significant workforce adaptation, creates economic value and enhanced employment opportunities rather than net job displacement (PWC, 2025).

An in-depth understanding of AI can be facilitated by examining its technical components, operational mechanisms, and associated terminology.

Table 1.1: ***Artificial Intelligence (AI): Definitions, Tools, and Terminology***

Category/Type	Definition
Artificial Intelligence	AI is the automation of cognitive processes. It uses advanced analytical techniques, such as machine learning, to collect, process, and interpret events to support and automate decision-making. In addition to machine learning (teaching computers to learn from data), it includes natural language processing (teaching computers to interpret and generate human language) and computer vision (teaching computers to interpret videos and images).
Generative AI (Gen AI)	Gen AI is a type of AI used for content creation that leverages existing data types such as text, video, and images. Examples include ChatGPT, Gemini, and Claude. Performance is based on training on vast amounts of text data to generate human-like responses.
AI Algorithms, Models, and Systems	An AI algorithm is a step-by-step procedure that a computer follows, which can lead to the creation of an AI model representing something that can be built using a combination of algorithms (Chartered Institute of Personnel and Development [CIPD], n.d.). A foundation model, or general-purpose AI system, results from Large Language Models (LLMs). ChatGPT, Gemini, and Claude are all built on LLMs and serve as examples of general-purpose AI systems. Not all Gen AI tools are general-purpose AI systems.

Source: Definitions adapted from Gartner (n.d.) and CIPD (n.d.).

Artificial Intelligence Barometer

The PWC (2025) Global Jobs Barometer analysed over one billion job advertisements across six continents to assess AI's global impact on employment, remuneration, skills requirements, and productivity. Key findings indicate that AI adoption is accelerating across all sectors, necessitating substantial job role redefinition. Notably, traditional credential requirements are evolving; some employers are reducing emphasis on formal degree qualifications for AI-related roles, instead prioritising demonstrated learning capacity and knowledge acquisition capabilities. This shift democratises access to knowledge-intensive employment. Preliminary evidence suggests that AI adoption is generating economic growth, though long-term impacts require continued empirical investigation.

Origin and Nature of Work

Work may be defined as the application of physical or cognitive effort directed toward producing goods, services, or outcomes of value to oneself or others. Over time, the meaning of work has evolved in response to social, economic, and technological changes. In pre-industrial societies, work was often tied to subsistence agriculture or craft-based production and was regarded as a subsistence necessity rather than a source of personal fulfilment. With the rise of industrialisation, work became increasingly formalised and specialised, with workers often performing repetitive tasks in factories or mines. This led to a growing sense of alienation and dissatisfaction among workers, as theorised by Marx (1867) and other social theorists. The post-war period witnessed a growing focus on job security and the significance of work as a source of personal identity and fulfilment. This was echoed in the rise of the welfare state systems, the development of social safety nets and workplace protections. However, economic volatility in the 1970s and 1980s reoriented organisational priorities toward efficiency and flexibility, leading to increased prevalence of part-time and temporary work arrangements. Contemporary scholarship increasingly addresses the gig economy, automation impacts, and lifelong learning imperatives as central to workforce development (Garritsen De Vries, 1986).

Work Characteristics Across Historical Eras

Table 1.2: *Characteristics of Work Across Historical Eras*

Era	Time Period	Characteristics
Prehistoric era	2.6 million years ago – 10,000 BCE	▪ Hunting, gathering, and nomadic lifestyles dominated. ▪ Primary focus was survival; work division was gender-based (males hunting, females gathering). ▪ Formal employment and career concepts did not exist; work met immediate subsistence needs.
Ancient era	3000 BCE – 476 CE	▪ Agriculture and animal domestication emerged, enabling settlement development. ▪ Work was stratified by social status, with enslaved peoples and servants performing arduous labour. ▪ Skilled workers (artisans, merchants) emerged; professions began to formalise.
Medieval era	476 CE – 1450 CE	▪ Feudalism dominated; work was tied to land and most populations were peasants. ▪ Guilds (associations of skilled workers) marked a significant shift in work organisation. ▪ Work increasingly became recognised as a means of generating livelihood beyond subsistence.
Industrial era	1760 CE – 1840 CE	▪ Factories and mass production emerged, creating novel occupational categories. ▪ Labour became highly specialised; workers performed discrete tasks in assembly-line production. ▪ Occupational roles expanded (factory workers, engineers, managers).
Modern era	1950 CE – present	▪ Transition from manufacturing to service-oriented economies occurred. ▪ Technological advances automated numerous tasks, creating employment in information technology, healthcare, and finance. • Work–life balance gained prominence; remote work expanded substantially, particularly following the COVID-19 pandemic.

Source: Developed by authors on the basis of historical labour market and occupational analysis.

Work involves exerting effort and cognitive capacity to produce goods or services of value to others. Work can take many forms, from paid employment to volunteer work, domestic labour, and artistic creation. The nature of work has evolved and varied across cultures and societies, reflecting different economic, political, and social systems.

An economic perspective views work as a primary means of securing livelihood. In contemporary capitalist societies, work is commodified, exchanged on labour markets, and individuals engage in paid employment to generate household income and accumulate wealth. This perspective emphasises work's instrumental function: income generation, capital creation, and economic growth.

Social and Cultural Dimensions

Beyond economic function, work holds profound social and cultural significance. It can be a source of identity, purpose, and meaning for individuals, as well as a means to contribute to society and build social connections. It can reflect cultural values, beliefs, and traditions and shape individual and collective identities. For example, some cultures highly value hard work and self-discipline, while others prioritise leisure and balance.

Psychological dimensions of work is essential to consider. Work can have a profound impact on an individual's mental and emotional well-being, as well as motivation, satisfaction, and performance. Factors such as job autonomy, social support, and work-life balance can influence employees' engagement and commitment to their work, as well as their overall happiness and quality of life. On the other hand, high levels of job stress, burnout, and dissatisfaction can contribute to various adverse outcomes, including poor mental health, reduced productivity, and turnover.

Macro-level factors, technological change, globalisation, and political power dynamics, fundamentally shape the nature of work. Automation and AI technologies are transforming employment structures, generating both opportunities and challenges for workers and organisations. Economic globalisation has enabled outsourcing, offshoring, and gig economy expansion. Labour policy, regulatory frameworks, and government interventions substantially influence work conditions, remuneration structures, and the distribution of employment-related benefits and risks.

Significant Definitions of Work

Significant scholarly definitions illuminate work's multifaceted nature. The following exemplars represent influential theoretical frameworks that continue to shape contemporary understanding of work and employment:

- Adam Smith (1776) conceptualised work as wealth generation through labour division, arguing that specialisation increases productivity and efficiency, thereby driving economic growth.
- Karl Marx (1867) analysed work as the generation of surplus value for capital accumulation, arguing that capitalist labour relations are inherently exploitative and necessitate structural transformation.
- Max Weber (1905) examined work as a means of achieving social status through wealth and power accumulation, arguing that Protestant ethical frameworks were instrumental in capitalism's emergence and that work is fundamental to modern society.
- Frederick Taylor (1911) pioneered scientific management, decomposing work into standardised, optimised tasks to maximise productivity and profit, an approach that fundamentally shaped industrial work organisation.
- Herbert Blumer (1969) conceptualised work as a negotiated, interpretive process, arguing that work's meaning and experience are shaped not solely by material conditions but by social and cultural dimensions.
- Harry Braverman (1974) analysed the deskilling of labour under capitalism, demonstrating how profit maximisation necessitates standardisation and the erosion of worker autonomy and creativity.
- Arlie Hochschild (1983) conceptualised emotional labour, the managed expression of emotion as a work requirement, as increasingly central to service-sector employment, with substantial emotional and psychological costs for workers.
- Emotional intelligence is defined as 'the ability to perceive and express emotion, understand and reason with emotion, and regulate emotion in oneself and others' (Mayer & Salovey, 1997, p. 10).

- Richard Sennett (1998) argued that work is foundational to personal identity and self-worth, and that the erosion of stable, meaningful employment negatively impacts individual and collective well-being.
- The entrepreneurial self is defined as 'a mode of subjectivity that emphasises the capacity for self-initiated action and the need to take responsibility for one's wellbeing' (Rose, 1999, p. 73).
- Gig work is defined as 'an activity in which an individual provides a service or performs a task, for remuneration, that does not create an employment relationship' (Kalleberg et al., 2000, p. 1), representing a substantial departure from traditional employment structures.
- The portfolio worker is characterised as one who 'actively manages a portfolio of paid and unpaid work, in which different elements of work, income, and personal development are mixed' (Sullivan & Arthur, 2006, p. 248).
- Guy Standing (2011) introduced the concept of the precariat, a class of workers characterised by insecurity, instability, and insufficient social protections, arguing that gig economy expansion has created substantial populations of workers lacking traditional employment security.
- Remote work is defined as 'any work done away from a central location or traditional office setting' (Golden et al., 2020, p. 2).
- Job crafting is defined as 'the process by which employees actively shape and mould their job characteristics, such as tasks, relationships, and meaning, to align with their personal goals and strengths' (Wrzesniewski & Dutton, 2021, p. 23).
- Digital nomadism is defined as 'a lifestyle in which individuals leverage digital technologies to earn a living whilst simultaneously travelling and residing in different locations worldwide' (Gregory & Valenti, 2021, p. 2).
- Algorithmic management is defined as 'the use of algorithms and data analytics to manage, control, and optimise work processes and employee performance' (Rosenblat & Stark, 2021, p. 2).

- Care is defined as 'the work of providing emotional, physical, and practical support to others in need, including children, elderly persons, and individuals with disabilities or chronic illnesses' (Thompson & Warhurst, 1998).

These definitions are important because they help us understand how work has been conceptualised and studied. By examining these different definitions, we can understand how work has been experienced, understood, and organised in other societies and historical periods. By considering these diverse definitions of work, we can gain a more nuanced and complex understanding of this complex phenomenon. It will help us understand how work has shaped human societies and cultures.

Perceptions of Work Across G7 Countries

Perceptions of work have shifted significantly across G7 countries over the last century, with transforming economic conditions, technological innovations, and social and political movements all playing a role. While each country has its distinct history and culture, shared patterns and common challenges are evident across all seven economies.

Figure 1.2: Per capita GDP based on PPP of G7 countries in 2023

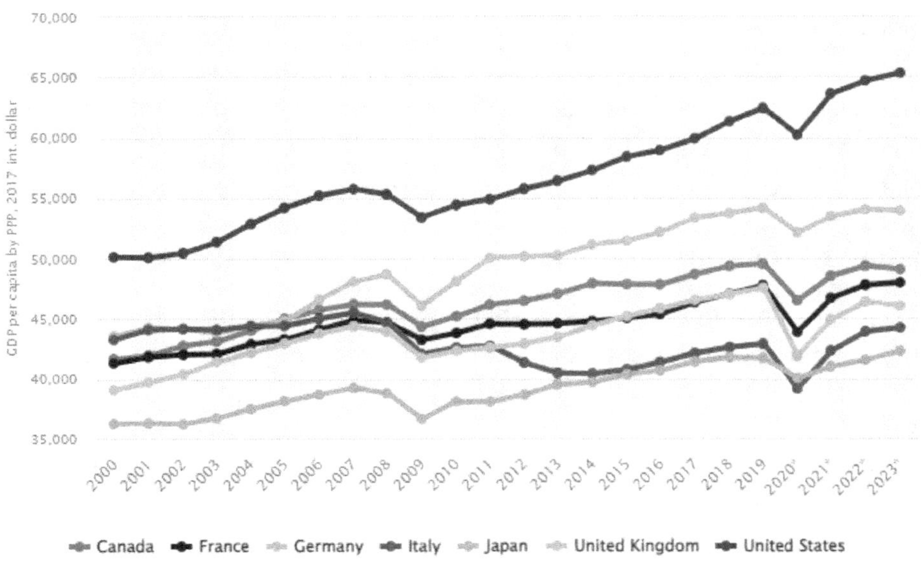

Source: www.statista.com

United States

The perception of work in the United States has shifted significantly over the last century, particularly regarding workers' rights and the role of government in the economy. Early 20th-century US work was characterised by harsh conditions and extended hours. Labour union emergence and progressive legislation, including the Fair Labour Standards Act and National Labour Relations Act, fundamentally altered labour relations. In the post-war period, not only did the US. experience economic growth and increasing prosperity, but also rising inequality and a declining middle class. The economy has become more service-oriented in recent decades, emphasising flexible, non-traditional work arrangements such as gig and remote work. Today, work is seen as a means of self-fulfilment and personal growth, emphasising the pursuit of meaningful and rewarding careers.

Canada

Canada has undergone significant changes in perceptions of work over the last century. In the early 1900s, the country had a strong tradition of long work hours, labour unions, poor working conditions, and low pay. However, with the rise of unions and the introduction of labour laws, the perception of work shifted to emphasise workers' rights, workplace safety, and social welfare policies, including universal healthcare and generous parental leave benefits. However, there has been a trend toward precarious work and rising income inequality, particularly in urban areas. In recent years, there has been a growing focus on issues such as the gender pay gap, the rights of Indigenous workers, flexible work arrangements, and the importance of diversity and inclusion.

United Kingdom

UK work perception has been shaped by substantial industrialisation history and trade unionism. In the early 20th century, workers' rights were advanced through legislation such as the Trade Disputes Act and the National Insurance Act. In the UK, work has traditionally been seen as a duty and a means of earning a living rather than a source of personal fulfilment. However, the post-war period saw the decline of traditional industries and the rise of a service-based economy, leading to increased casualisation and job insecurity for many workers. The shift in the economy has led to changes in job types, the labour market, and skills and qualifications. Recently, there has been a growing focus on the gig economy, remote work, self-employment, zero-hours contracts, and stronger worker protections. As in Canada, there has been growing recognition of the importance of well-being and work-life balance in the UK.

France

France has a strong tradition of labour activism and workers' rights, dating back to the 19th century. Unions traditionally had a strong influence on the perception of work. The country has a comprehensive social safety net, including universal healthcare and generous unemployment benefits. In France, work has traditionally been seen as a means of social mobility, and there is a strong emphasis on education and professional achievement. In 1981, the government implemented the 35-hour workweek, which reduced the time workers were expected to spend at the job. Traditionally, women did not have access to male-dominated professions, but now diversity and inclusion have become increasingly important. There is a strong emphasis on education and professional achievement. However, there is also a cultural value placed on leisure time and work-life balance, and many French workers prioritise the quality of life over career advancement.

Germany

Germany is recognised for workforce organisation and efficiency, substantially enabled by trade union and works council participation in labour relations. Historically, German work perception emphasises duty as a means of contributing to society, with a strong emphasis on technical education, vocational training, and apprenticeships, which has helped to promote a skilled and adaptable workforce. Work councils, like unions, are elected bodies that represent company employees and strongly influence perceptions of work in Germany. Traditionally, job security is highly valued, with many Germans seeking long-term employment with a single employer, but now workers are embracing non-traditional forms of employment. However, a cultural value is also placed on work-life balance, and German workers enjoy some of the shortest working hours in Europe. In recent years, there has been growing concern about the impact of globalisation and automation on traditional manufacturing jobs and a focus on promoting specialised knowledge, expertise, innovation, and digital skills.

Italy

During the 1960s and 1970s economic boom, Italian work was central to identity and life experience. Italians could secure stable, well-paying jobs in manufacturing, construction, and other sectors. Work was seen as a way to build a better life for oneself and one's family, and unions played a prominent role in advocating for workers' rights. However, the economic crises of the 1980s and 1990s brought about significant changes in Italy's perception of work. High unemployment rates, particularly among young people, led to disillusionment and frustration with the labour market. Many Italians began to view work as something to be endured rather than enjoyed, and job security became increasingly elusive. In recent years, perceptions of work in Italy have been shaped by factors such as the rise of the gig economy and the growth of the service sector. Many Italians value job security and stability, but there is also a growing recognition of the importance of work-life balance and the need for flexibility and autonomy in the workplace.

Japan

Post-World War II, Japan's economy experienced rapid growth, and many Japanese workers viewed their jobs as a source of stability and security. Lifetime employment at a single company was the norm in exchange for job security and benefits. During the 1980s, Japan's economy continued to boom, and the country became known for its "salaryman" culture, with long working hours and a strong emphasis on company loyalty. However, the bubble economy burst in the early 1990s, leading to economic stagnation and high unemployment. This period was marked by a new class of workers, including part-time and contract workers, who lacked the job security and benefits of traditional full-time employees. In the late 1990s and early 2000s, Japan began to grapple with issues related to work-life balance, as the long hours and high-pressure work environment were linked to a rise in mental health problems and a decline in birth rates. The government introduced policies to encourage better work-life balance, including limits on overtime hours, promotion of telecommuting, and a flexible, inclusive, and sustainable work environment that supports the well-being of workers and the broader society.

Work perception across G7 nations has evolved significantly over the last century, with changing economic conditions, technological innovations, and social and political movements all playing a role. While each country has its distinct history and culture, many shared trends and challenges reflect broader global shifts. Work perception remains dynamic, constantly responding to emerging political, economic, social, and technological trends and worldwide developments.

The Book's Approach and Design

This book employs qualitative research methodologies to generate a rich, in-depth understanding of the future of work across G7 nations. Qualitative research prioritises data saturation, the point at which additional data collection yields no substantively new insights, rather than predetermined sample sizes. The number of participants required to achieve saturation varies according to research complexity, phenomenon multidimensionality, and desired analytical depth.

Therefore, the sample size in qualitative research is often determined iteratively, with researchers continuously assessing the data during the research process. Researchers typically continue collecting data until reaching saturation, at which point further data collection does not yield significant new insights or themes.

Qualitative research methodology prioritises depth and richness of the data rather than the quantity of participants. Researchers prioritise the quality of information and the richness of participants' experiences and perspectives to comprehensively understand the research topic.

Book Structure

The book comprises three principal sections. The first begins with an introduction to the future of work. Then, the book takes the reader on a journey about the origin and nature of work. This section examines perceptions of work in the G7 countries. The authors then describe one of the biggest disruptors of all time, COVID-19. Then, the authors detail hybrid work strategies and highlight employee empowerment initiatives implemented across various countries and companies.

The second part of the book discusses the future of the work landscape in each of the G7 countries. The authors delve into possibilities, realities, and challenges facing each country. Part two of the book portrays the changes in the coming years, the influence of technology on the labour market, the shifting perceptions of work across countries, and the roles of governments, corporations, and individuals in shaping the future of work.

The third part of the book reveals the analysis of twenty-two futurists of work, company heads, professors, HR professionals, entrepreneurs, editors, thought leaders, strategists, labour market experts, administrators, veterans, and professors to obtain in-depth knowledge about the changes in the future of work and its implications for the labour market. The authors have adopted the STEEPV frameworks to analyse the interview transcripts and share expert insights to identify specific research themes that will shed light on the dominant characteristics of the future of work. The authors have analysed data to offer country-based recommendations.

Chapter 2

Changes in Work Patterns

Introduction

This chapter examines the principal drivers of change in work patterns across the G7, focusing on COVID-19 as a catalytic disruption, the subsequent adoption of hybrid work models, and emerging organisational responses to contemporary workforce expectations. The analysis establishes the foundation for understanding the technological and social shifts that define contemporary work, through country-level analysis and expert perspectives.

COVID-19: The Major Disruptor

Global Scale and Societal Impact

The 21st century has experienced numerous significant disruptions, including the 2008 financial crisis, the proliferation of social media and platform economies, the ongoing climate crisis, and the COVID-19 pandemic (first identified in late 2019). Among these, the COVID-19 pandemic ranks among the most consequential disruptions in scale and global scope (Kolluru et al, 2021; World Health Organisation, 2021).

The pandemic caused widespread illness, death, and unprecedented economic hardship across all sectors. Between 2020 and 2021, the virus infected hundreds of millions globally, disrupted international trade and travel, triggered supply chain collapse, and forced governments to impose lockdown measures affecting billions of workers (Kolluru et al, 2021; United Nations Development Programme, 2021). Beyond immediate health impacts, COVID-19 exposed and amplified existing structural inequalities: the pandemic disproportionately affected marginalised communities, low-income workers, people of colour, and immigrants who faced heightened occupational and economic risk due to underlying social and economic disparities (Bambra et al., 2020; International Labour Organisation, 2021).

Sectoral Impacts on Employment and Operations

The pandemic's workplace disruptions manifested distinctly across sectors, each shaped by occupational exposure, supply chain dependency, and service delivery models:

- **Healthcare:** Workers faced dual pressures: extended working hours responding to surging patient demand, combined with critical shortages of personal protective equipment (PPE) that significantly increased occupational infection risk (Lancet Commission on the Future of Work, 2022). This pressure cascaded beyond direct clinical staff to ancillary and administrative functions.

- **Education:** Widespread school and university closures forced a rapid transition to remote learning, disrupting students' academic progress and imposing childcare and work-balance challenges for parents and educators navigating simultaneous professional and caregiving responsibilities (UNESCO, 2021).

- **Hospitality and Tourism:** The hospitality sector, including restaurants, hotels, and travel services, experienced severe contraction as travel restrictions curtailed demand and forced temporary closures. Revenue collapsed sharply in 2020, with cascading job losses concentrated among low-wage service workers, many of whom lacked employment protections or benefit eligibility (International Labour Organisation, 2020).

- **Manufacturing:** Manufacturing faced supply chain fragmentation and reduced product demand stemming from locked-down consumer economies. Factory closures during the 2020–2021 initial phases reduced production capacity; subsequent social distancing requirements and enhanced hygiene protocols increased operational costs and further slowed output (World Economic Forum, 2021).

- **Retail:** Retail closures accelerated the structural shift toward e-commerce and online shopping, paradoxically increasing demand for warehouse and logistics employment even as traditional retail jobs contracted (McKinsey & Company, 2021). This sectoral divergence created significant labour market dislocations.

- **Agriculture:** Labour shortages caused by travel restrictions and border closures constrained agricultural harvesting and livestock management. Combined with supply chain disruptions and reduced demand for certain crops, farmers' incomes suffered alongside increased food price volatility in some regions (Food and Agriculture Organisation of the United Nations, 2021).

COVID-19 and the Acceleration of Work Transformation

The pandemic fundamentally altered work patterns and organisational strategy in two critical ways. First, it accelerated existing trends: businesses, workers, and professionals adapted to new working conditions, including hybrid arrangements and digital collaboration technologies that had existed in nascent form pre-2020 (Bloom et al., 2024). Second, it brought forward organisational and social shifts that might otherwise have required years to implement, compressing transformation timelines into months.

The emerging work landscape presented both opportunities and challenges. Positive dimensions included flexible work arrangements enabling improved work-life integration, reduced commuting time and associated burnout, and for many workers, increased productivity facilitated by personalised work environments. However, the pandemic simultaneously exposed critical inequalities: widespread job insecurity, income inequality intensified by sectoral divergence, unequal access to technology and remote working capability, and the emergence of mental health and social isolation concerns among remote workers (International Labour Organisation, 2021). These contradictions define the future-of-work landscape, requiring nuanced policy and organisational responses.

Changing Work Strategies: The Hybrid Work Model

Emergence and Pre-Pandemic Foundation

Hybrid work arrangements, variously termed 'blended work' or 'flexible work arrangements', refer to employment structures allowing workers to divide time between office-based and remote settings. This arrangement emerged from three converging factors: technological capability enabling cloud-based collaboration, employee demand for work-life integration, and organisational cost management through reduced real estate footprint (Bloom et al., 2024).

Critically, many corporations globally had adopted hybrid structures before the COVID-19 pandemic. Pre-pandemic literature clearly demonstrates that the pandemic forced most organisations to adopt or significantly expand hybrid models to survive and sustain operations amid lockdowns and government-imposed restrictions (Harvard Business School, 2021).

Research Evidence on Hybrid Work Effectiveness

Numerous studies substantiate that remote and hybrid work can be both effective and productive. Research from leading institutions indicates that organisational practices emphasising delegation, autonomy, and flexibility correlate with higher employee engagement than traditional micromanagement approaches (Bloom et al., 2024). Contemporary office design, featuring spaces facilitating concentration for complex, focused work, increasingly characterises the hybrid workplace, reflecting evidence that physical environment design significantly influences productivity and well-being (Cushman & Wakefield, 2021).

Critically, evidence-based organisational research on hybrid effectiveness emphasises that employee autonomy correlates with improved outcomes. Employees demonstrate higher productivity and engagement when given decision-making authority regarding work location and schedule, provided performance metrics are clearly defined and regularly assessed (Bloom et al., 2024). This shift reflects the evolving management philosophy required for distributed, asynchronous work environments.

Types of Hybrid Work Models

Organisations adopt diverse hybrid models aligned with their specific operational, sector, and strategic needs. Table 2.1 presents eight prevalent hybrid work configurations:

Table 2.1: Popular Forms of Hybrid Work Models

Type of Hybrid Model	Description
ROWE (Results-Only Work Environment)	Employees are evaluated exclusively on outcomes; freedom to determine when, where, and how designated tasks are completed, with Key Performance Indicators (KPIs) as the sole measure of productivity.
Designated team models	Mixed approach where some teams work office-based whilst others operate remotely, determined by role and team function.
Split week models	Employees divide working time between office-based days and home-based days on a regular schedule (e.g. three days office, two days remote).
Rotating schedules	Teams rotate between remote and onsite work on weekly or fortnightly cycles, allowing shared office resources and scheduled in-person collaboration.
Flexibility arrangement	Employees retain discretion to choose when and where to work, subject to job-specific requirements and team coordination needs.
Core hours	Employees must be present during defined core hours (e.g., 10:00–15:00) but work flexibly elsewhere outside these periods.
Project-based work	Teams work remotely on specific, bounded projects with occasional scheduled in-person meetings and collaboration sessions.
Office-first model	Employees typically work from the office (4–5 days weekly) with limited remote options (1–2 days weekly or several days monthly) at discretion.

Source: ROWE = Results-Only Work Environment; KPI = Key Performance Indicator. Developed by the authors.

Implementation Challenges and Requirements

Hybrid work requires rigorous management to prevent organisational and employee deterioration. The UK 'Return to Work' research report found that 60% of employees missed informal collegial interactions, highlighting a critical challenge to hybrid effectiveness (Return to Work Research, 2021). Rapid role transition to hybrid arrangements imposes cognitive and emotional demands on employees. This transition depletes emotional and cognitive resources, potentially exacerbating burnout risk (Lancet Commission on the Future of Work, 2022). Unless organisations provide structured support managing distributed reporting relationships, time zone complexity, and rapidly introduced technologies, transition stress may undermine the productivity gains attributed to flexible work models.

The Way Forward: Enabling Effective Hybrid Working

Organisational Infrastructure and Workspace Design

The upcoming hybrid workplace prioritises collaboration, brainstorming, and collective problem-solving, requiring contemporary design featuring large, dedicated collaborative workspaces. Evidence-based workspace design incorporates open visibility zones that stimulate value-generating conversations and interactions among colleagues (Cushman & Wakefield, 2021). This approach contrasts with pre-pandemic high-density seating models.

Flexible workspace arrangements offer tangible operational and financial benefits: reduced real estate costs, smaller organisational footprints, and enhanced resource allocation. Organisations can create dedicated spaces for energetic, collaborative work, 'war rooms' for intensive projects, open workstations for routine tasks, and purpose-built meeting facilities for virtual and hybrid meetings, bookable according to team requirements and project phases (Harvard Business School, 2021).

Communication Infrastructure and Cultural Foundations

Evidence-based communication infrastructure proves critical for hybrid success. Organisations must establish clear communication protocols specifying response time expectations, preferred communication channels for different purposes, and standardised procedures for virtual meetings, including accessibility and participation safeguards (Bloom et al., 2024). These protocols reduce role transition stress and improve decision-making speed in distributed teams.

Complementing the communication structure, organisational culture must foster meaningful, value-creating collaboration. The foundational value underpinning effective hybrid work is trust. A culture of trust empowers employees to manage asynchronous work, take initiative without constant oversight, and maintain engagement despite physical distance (Harvard Business School, 2021). Such trust-based cultures correlate with higher retention and reduced burnout (Bloom et al., 2024).

Evidence from Leading Organisations

- **Stanford Research Perspective:** Nicholas Bloom, a professor at Stanford Graduate School of Business who has studied hybrid work extensively, reports that hybrid arrangements with typical office presence days are sustainable, with research suggesting minimal to no documented impact on productivity, culture, team connectivity, or innovation (Bloom et al., 2024).
- **London Business School Perspective:** Linda Gratton, professor at the London Business School and author of *Redesigning Work*, argues that organisations should focus on the nature of work tasks and the outcomes they serve, rather than mandating uniform work location policies (Gratton, 2022). She categorises work into three types: (1) *focused work*, individual concentration tasks such as writing reports or data analysis; (2) *coordinated work*, project management requiring sequential interdependence; and (3) *cooperative work*, brainstorming and collaborative tasks requiring real-time interaction (Gratton, 2022). This taxonomy suggests that different work types benefit from different location strategies: focused work suits remote environments, whilst cooperative work benefits from in-person settings.
- **HubSpot Case Study:** Even HubSpot's Chief People Officer, Katie Barke, emphasises that providing employees with flexibility is critical to organisational performance. HubSpot, with over 5,500 staff globally, has adopted the "3E Strategy": (1) *enablement* is providing managers and staff with tools supporting hybrid meetings and team connectivity; (2) *experimentation* is conducting virtual events and continuous learning programmes to improve hybrid practices; and (3) *empathy* is fostering a culture of support, offering exceptional training, and encouraging employees to share experiences and challenges (HubSpot, 2023).
- **Microsoft Implementation:** Microsoft allows employees to work from home up to 50% of the time within a flexible scheduling framework. However, policy changes effective early 2026 require most employees living within 50 miles of a Microsoft office to work onsite a minimum of three days per week, reflecting organisational prioritisation of in-person collaboration and innovation (Microsoft, 2025).

- **Shopify's Remote-First Model:** In contrast, Shopify has embraced a fully remote operating model, with all employees working from home whilst maintaining close communication and collaboration through scheduled virtual meetings and corporate events. Shopify has designed a digital 'culture deck' enabling employees to feel connected to the company's mission and values despite geographic distribution (Shopify, 2024). Recent changes to benefits (ending internet reimbursement and certain professional development credits), have not shifted the remote policy back to in-office requirements, signalling sustained organisational commitment to remote working.

- **Spotify's Work-from-Anywhere Policy:** Spotify's organisational philosophy, encapsulated as "Work is not a place you come to but what you do," reinforces its work-from-anywhere policy. Since February 2021, 6,500 Spotify employees have held the option to work from home, from a mix of office and home, or from locations without a Spotify hub nearby (Spotify, 2023). The organisation pays for coworking space access for distributed employees and has updated compensation structures to reflect the country of work rather than the city of employment, acknowledging cost-of-living variation globally.

  The policy's outcomes demonstrate tangible diversity gains within 18 months, 2% of staff relocated to different countries and 4% moved within the United States. Women in leadership positions increased from 25% (2019) to 48% (2021), and Black and Hispanic representation increased from 12.7% to 18%, suggesting that geographic flexibility expands the talent pool (Spotify, 2023). These shifts reflect the principle that work-from-anywhere policies can enhance organisational diversity by removing geographic barriers to opportunity.

- **Dropbox Flexible Arrangement Model:** Dropbox has implemented a flexible work model allowing employees to determine when and where they work, contingent upon meeting performance goals. The organisation has redesigned office spaces to prioritise collaboration and teamwork, featuring open workstations and dedicated collaborative rooms for brainstorming and knowledge-sharing (Dropbox, 2024). As of 2025, Dropbox maintains its flexible, virtual-first approach, granting employees autonomy in work location and schedule decisions, provided performance metrics are achieved.

Synthesis: Organisational Diversity in Hybrid Implementation

The case studies above illustrate that leading multinational organisations pursue diverse, contextually appropriate hybrid strategies. Some prioritise in-office presence (Microsoft, Goldman Sachs), others embrace remote-first models (Shopify, Spotify), and many adopt differentiated approaches by role and team function (Salesforce, General Electric). Importantly, no single hybrid model emerges as universally optimal; rather, effectiveness correlates with transparent communication, clearly defined performance expectations, and organisational cultures that trust employee autonomy (Bloom et al., 2024). By prioritising flexibility, collaboration infrastructure, and effective communication, these organisations have created work environments enabling employee effectiveness regardless of physical location.

Employee Empowerment Initiatives Across the World

Four-Day Work Week and Compressed Working Hours

Post-COVID-19, the world is experiencing significant experimentation with work schedule restructuring. Numerous organisations worldwide have implemented employee empowerment initiatives, particularly shortened working weeks, to respond to unprecedented change and ensure employee well-being. The four-day work week concept, though dating back decades in academic discussion, has undergone a transformative redefinition in recent implementations, offering greater flexibility and enhanced time off whilst maintaining compensation and well-being outcomes (Kellogg Foundation, 2022).

United Kingdom and Ireland

Two hundred organisations in the UK have adopted the four-day work week, with Scottish employees saying, *"We will never go back to the traditional work-from-office approach."* (Institute for the Future of Work, 2023). These adoptions implemented without loss of pay across the UK and Ireland demonstrate that shortened working weeks can maintain economic viability. Governments and organisations implementing such initiatives have reported encouraging outcomes supporting work-from-home adoption and employee retention.

Belgium

Belgium has announced that thousands of federal civil servants will no longer be required to respond to work communications outside designated working hours, establishing a statutory 'right to disconnect.' Concurrently, around 6,500 government workers earned entitlement to a four-day work week. The Belgian government's rationale posits that this employee empowerment policy will enhance economic innovation and digital transformation capability, strengthen individuals' well-being, and improve organisational performance (Belgian Federal Government, 2023).

Scotland

Scotland has conducted extensive trials with working hour reductions of 20%, enabling workers to access four-day weeks without compensation loss. Scottish businesses have already introduced varied compressed working week models adapted to sector-specific requirements (Scottish Government, 2023).

Spain and Iceland

Spain is piloting a 32-hour working week over a three-year experimental period without reducing worker compensation, a model distinct from Belgium's 38-hour compressed week (Spanish Government, 2023). Iceland conducted longitudinal research from 2015 to 2019 covering 2,500 workers (representing 1% of the national workforce), assessing shortened working days' impact on productivity, employee well-being, and work-life balance. Results demonstrated significant positive outcomes: worker stress and burnout declined measurably, work-life balance improved substantially, and productivity remained stable or increased (Iceland Technology Authority, 2021). Based on these findings, nearly 90% of Iceland's working population now access either reduced hours or alternative work patterns among the highest adoption rates globally. Europe and beyond. Countries experimenting with four-day working weeks extend beyond early adopters and include Australia, Canada, Denmark, Finland, Germany, Ireland, Lithuania, the Netherlands, New Zealand, Sweden, and the United States. Portugal has introduced comprehensive labour law changes, including statutory prohibitions on employer contact with staff outside working hours and the prohibition of workplace monitoring for home-based employees; firms must also reimburse technology and utility expenses (electricity, internet) incurred by remote workers (Portuguese Parliament, 2023). These protections reflect legislative recognition that remote working introduces new equity and well-being considerations.

Japan and the United Arab Emirates

Japan, culturally characterised by 'live to work' professional norms, is exploring four-day working week adoption. Microsoft Japan conducted an experiment in summer 2019 called the 'Work-Life Choice Challenge,' enabling 2,300 employees to participate in shortened working week options. Results indicated that workers were significantly happier and 40% more productive during the trial period, suggesting that cultural norms around excessive working hours can shift with structured experimentation (Microsoft Japan, 2020). The United Arab Emirates has adopted a four-day federal workweek with weekends commencing at midday on Friday and extending through Sunday, representing a significant regional policy shift.

Organisational Responses: Case Studies in Implementation

Google

Google adopted future-of-work initiatives during the COVID-19 pandemic, offering flexible work schedules, remote work options, and innovative workspaces designed for both focus and collaboration. The organisation invested heavily in automation, artificial intelligence, and machine learning to enhance productivity and efficiency. However, Google has placed limits on its 'work-from-anywhere' allowance, restricting annual access to four calendar weeks. Significantly, even a single day of remote work counts as a full week toward this annual allowance, and working from different Google offices (in different states or countries) is not permitted. Google requires employees to live within 50 miles of their assigned office and maintain on-site presence a minimum of three days weekly; the organisation tracks attendance and incorporates it into performance appraisals, with non-compliance triggering disciplinary action (Google, 2024).

Amazon

Amazon has emerged as a leader in automation and AI adoption, integrating robotic automation, machine learning, and artificial intelligence to streamline operations. During the COVID-19 period, the organisation embraced remote work and flexible scheduling. However, Amazon has mandated that employees return to full-time onsite work (five days weekly), with organisational leadership contending that in-person interaction promotes collaboration and innovation (Amazon, 2024). Simultaneously, Amazon has positioned itself at the forefront of workforce upskilling, investing over $2.5 billion to prepare 50 million people for future-of-work transitions. To date, 700,000 employees have completed upskilling through Amazon's paid education and training programmes, with 425,000 from the United States, exceeding 2019 organisational targets by more than sevenfold (Sadie, 2025).

Salesforce

Salesforce offers flexible work schedules and remote work options, implementing a range of collaboration tools to enhance communication and productivity. The organisation continues operating under a hybrid, flexible model with options including 'office-flex' and 'Return & Remote' arrangements tailored to team requirements and roles (Salesforce, 2024). This approach allows teams to select their preferred office-to-remote work mix. Office redesigns prioritise collaboration, innovation, and scheduled team in-person events, balancing distributed work with collective engagement.

General Electric (GE)

GE has embraced automation, artificial intelligence, and machine learning for operational streamlining. The organisation offers hybrid work models differentiated by division and location, providing hybrid scheduling and full-time remote arrangements subject to team requirements and management approval (GE, 2024). GE has invested in developing a workforce skilled in emerging technologies, particularly artificial intelligence, digital analytics, and advanced manufacturing, enabling adaptation to rapid technological change.

Toyota

Toyota, a leading automobile manufacturer, has adopted automation, robotics, and artificial intelligence to enhance efficiency and productivity. The organisation has introduced flexible work schedules and remote options to support employee work-life balance. Reflecting evolving workforce expectations, Toyota modified its policy effective 2025, now requiring onsite presence four days weekly, positioning the manufacturer between traditional and hybrid models.

Hilton, Nestlé, and Goldman Sachs

Hilton has introduced innovative customer-facing technologies (mobile check-in, digital room keys, mobile ordering) and flexible work options for back-office staff. Nestlé has similarly adopted flexible and remote work arrangements supporting work-life balance. Goldman Sachs offers flexible schedules, remote options, and collaborative workspace designs, investing substantially in workforce development for technological adaptation. However, effective March 2025, Goldman Sachs mandated full-time onsite work (five days weekly), signalling a return-to-office priority for this financial services sector leader (Goldman Sachs, 2025).

Automation and Artificial Intelligence: Current Challenges Facing the Future of Work

The future of work faces several significant challenges requiring proactive policy and organisational responses. Automation and artificial intelligence (AI) are substantially disrupting labour markets, skill requirements, and employment structures. Understanding these challenges is essential for developing evidence-based policy and organisational strategy.

Scale of Job Displacement and Transition

Automation and AI present dual labour market impacts: job displacement in routine occupations and job creation in emerging fields. According to the World Economic Forum (2025), as many as 92 million jobs may be displaced by automation by 2030, whilst 170 million new roles may emerge, a net displacement when accounting for skills mismatch and geographic variation.

This shift necessitates new approaches to education and training and policy frameworks ensuring inclusive workforce transitions.

The skills challenge compounds the displacement risk. Many future-of-work roles demand advanced competencies, particularly in technology, healthcare, and engineering. However, the global workforce faces a significant skills gap: approximately 40% of employers report difficulty filling job vacancies due to insufficient qualified candidates (International Labour Organisation, 2019). Concurrently, 40% of employers anticipate workforce reductions due to increased automation and AI implementation (World Economic Forum, 2025). Addressing this gap requires concerted investment in education and training programmes providing workers with skills necessary for future employment, alongside policy frameworks supporting workforce transitions and income security during retraining periods.

The Gig Economy: Opportunities and Structural Inequities

The gig economy, encompassing platforms such as Uber, Lyft, and Airbnb, has expanded substantially in recent years, creating new income opportunities for workers whilst simultaneously introducing new structural inequities. Gig work offers flexibility but lacks the employment protections and benefit structures characterising traditional employment: many gig workers lack access to healthcare, retirement plans, and paid leave. Income volatility, absence of employer-sponsored benefits, and weak bargaining power create conditions wherein workers struggle to meet basic needs and face heightened economic insecurity (International Labour Organisation, 2021). These structural gaps have generated calls for regulatory frameworks protecting gig workers and ensuring access to essential benefits and protections.

Diversity, Inclusion, and Equity in the Workplace

Despite documented progress, many workplaces continue addressing diversity and inclusion challenges. Persistent discrimination and bias against marginalised groups limit job opportunities and career advancement, impacting labour market participation and long-term earnings potential. Research by McKinsey & Company (2020) demonstrates that organisations with diverse workforces outperform less diverse peers, establishing a business rationale for equitable hiring and advancement. Addressing workplace diversity and inclusion requires sustained organisational commitment to creating genuinely equitable, inclusive work environments, coupled with accountability mechanisms and leadership sponsorship of equity initiatives.

Climate Change, Sectoral Transitions, and Workforce Adaptation

Climate change will substantially reshape labour market structures in the coming decades. Industries dependent on natural systems, such as agriculture, forestry and fisheries, can face significant disruption from altered weather patterns, increased extreme weather events, and ecosystem changes. Simultaneously, substantial employment growth is anticipated in renewable energy, sustainable agriculture, and environmental remediation sectors. Workforce transitions from carbon-intensive to sustainable industries require coordinated investment in education, training, and regional economic development, coupled with policies supporting affected workers and communities during sectoral transitions (International Labour Organisation, 2021).

Policy and Organisational Imperatives

Addressing these multifaceted challenges requires coordinated action among governments, employers, educational institutions, and workers. Key imperatives include: (1) investment in lifelong learning and workforce development programmes enabling skills transition; (2) policy frameworks establishing social protection for gig and contingent workers; (3) organisational and policy commitment to workplace diversity and equity; and (4) strategic planning for climate-driven sectoral transitions and green employment growth. The future of work depends on ensuring that technological change, economic restructuring, and environmental adaptation benefit workers equitably rather than concentrating gains amongst high-skill populations whilst displacing vulnerable workers.

Automation's Effect on the Future of Work

Scope and Scale of Automation Adoption

Automation is expected to significantly reshape work roles, skill requirements, wages, working conditions, and social structures over the coming decade. Advances in robotics, artificial intelligence, and algorithmic automation offer potential benefits such as increased productivity, efficiency, and innovation, but simultaneously raise critical concerns regarding job displacement, inequality amplification, and ethical implications of algorithmic decision-making.

Many sectors face accelerating automation. Manufacturing represents the earliest and most heavily automated industrial sector; robots perform assembly, welding, painting, and quality control functions. The International Federation of Robotics (2025) estimates that 4 million new industrial robots will be installed globally, continuing this trajectory. Transportation faces potentially transformative automation: autonomous vehicles are undergoing extensive testing and are expected to proliferate, potentially displacing millions of drivers and related transportation workers (World Economic Forum, 2018). Retail has experienced substantial automation through self-checkout systems, automated warehouses, and delivery drone pilots, displacing cashiers, stockers, and retail workers whilst simultaneously creating demand for warehouse automation specialists and logistics coordinators (Manyika et al., 2017). Finance has undergone an algorithmic transformation: robotic advisors provide investment guidance, and algorithms manage trading and risk functions, displacing some financial advisory roles whilst creating new roles in quantitative analysis and algorithmic development (Manyika et al., 2017). Healthcare increasingly employs robots for surgical assistance and patient support, and artificial intelligence increasingly analyses medical data for diagnosis support, potentially displacing some administrative and routine diagnostic roles whilst improving efficiency and clinical outcomes (Manyika et al., 2017).

The extent and pace of automation impact vary significantly depending on the specific technologies involved, the nature of work being automated, and the regulatory and social contexts in which automation is implemented. Some sectors may experience rapid displacement (transportation, routine retail), whilst others may see gradual workforce evolution (healthcare, finance).

Job Redefinition: From Task Replacement to Role Transformation

Rather than resulting in simple job elimination, automation is likely to redefine job roles and responsibilities. Machines increasingly assume routine and repetitive tasks such as data entry, basic accounting, simple customer service, freeing human workers to focus on higher-level cognitive work requiring creativity, complex problem-solving, emotional intelligence, and interpersonal skills (Manyika et al., 2017). This shift creates potential for more engaging, intellectually stimulating work, provided workers can transition into these roles. However, this potential is contingent on adequate training investment and educational infrastructure.

Skill Requirements and Education Transformation

Automation fundamentally alters workforce skill demands. Technical skills such as programming, data analysis, AI/machine learning operations, and cybersecurity increasingly differentiate employment opportunities. Simultaneously, soft skills like communication, collaboration, adaptive capacity, and emotional intelligence become increasingly critical for human-centric roles that machines cannot efficiently replicate (World Economic Forum, 2018). Workers must engage in continuous upskilling and reskilling throughout their careers, requiring systems of accessible, affordable lifelong learning. Traditional education systems designed for initial workforce preparation inadequately address this need, necessitating innovative partnerships between employers, educational providers, and government supporting career-long skill development.

Wages, Working Conditions, and Labour Market Inequality

Automation's impact on wages and working conditions remains contested. Some research suggests automation increases productivity and can support wage growth in affected sectors; others warn that automation increases inequality and joblessness (Manyika et al., 2017). A critical concern is the potential for a 'two-tier' labour market wherein high-skilled workers benefit from elevated wages and job security whilst low-skilled workers face displacement and economic insecurity (World Economic Forum, 2018). Without proactive policy intervention, including job transition support, income protection during retraining, and investment in accessible skill development, automation threatens to exacerbate wage inequality and labour market segmentation.

Ethical and Social Implications

Automation raises substantial ethical and social concerns requiring deliberate governance frameworks. Key concerns include: (1) worker rights and job security as automation reduces labour demand; (2) algorithmic bias in automated decision-making systems used for hiring, performance evaluation, and termination; (3) impact on worker surveillance and autonomy as algorithmic management systems proliferate; and (4) broader social implications regarding income distribution, social cohesion, and traditional family structures dependent on full-time employment (OECD, 2019). These considerations demand that automation policy integrate ethical frameworks and worker protection mechanisms, not merely productivity optimisation.

Navigating Automation: Integrated Policy Framework

The impact of automation on the future of work is substantial and wide-ranging. Whilst potential benefits include increased productivity and efficiency, legitimate concerns regarding job displacement, skill requirement changes, and labour market inequality require proactive governance.

Navigating automation requires coordinated action from policymakers, employers, and workers:

1. **Education and Skills Investment:** Invest substantially in lifelong learning systems providing workers with advanced technical skills (programming, data analysis, AI operations) and essential soft skills (communication, collaboration, creativity, adaptability). This requires institutional innovation beyond traditional education models.

2. **Worker Transition Support:** Develop comprehensive policies and programmes supporting workers displaced by automation, including income support during retraining, accessible skill development, and regional economic development strategies enabling sectoral transitions.

3. **Regulatory Frameworks:** Develop governance frameworks promoting responsible and ethical automation, addressing algorithmic transparency, bias mitigation, worker surveillance limits, and worker participation in automation decisions affecting their roles.

4. **Labour Market Protections:** Establish or strengthen worker protections, including bargaining rights, income security, benefit portability across employment forms, and enforcement mechanisms, ensuring that automation benefits are distributed equitably and workers retain dignity and security throughout transitions.

The future of work depends on whether technological change serves or concentrates gains amongst capital owners and high-skill workers. Proactive, comprehensive policy integration is essential to ensuring equitable, inclusive automation.

Chapter 3

Future of Work in the G7 – United States

Introduction

The United States (US), with a population of 349 million (Figure 3.2) (Wordometer, 2026), represents the largest economy within the G7. Approximately 96 million of the US population reside in 315 cities across 3,000 counties, including 25 megacities associated with high development, growth, and substantial job generation since the Great Recession (Lund et al., 2019). Conversely, 54 cities and 2,000 rural counties are home to older populations with lower educational attainment, high unemployment, and shrinking workforces. Between these extremes lie thriving cities with a large, mixed middle class experiencing modest economic growth (Lund et al., 2019). These geographic disparities suggest significantly different stages of automation adoption across future of work (FOW) scenarios.

The 25 megacities are predicted to reach 60% automation by 2030, whilst the remainder of the country lags. Within these scenarios, the impact of automation on Black American communities is of particular concern, as the racial wealth gap is projected to widen (Cook et al., 2019; Lund et al., 2019).

Figure 3.1: New York City and the Statue of Liberty

Source: Getty Images

Figure 3.2 shows US population trends with an overall increase in population for the period up to 2100

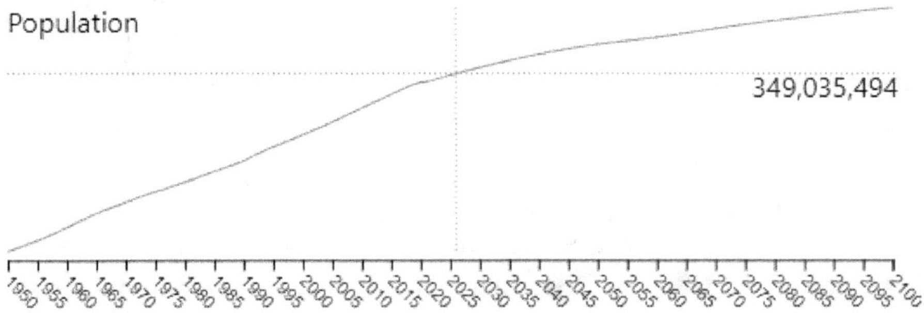

Source: Populationpyramid.net (2026)

Prior to the COVID-19 pandemic, predictions regarding the future of work for this vast economy focused on the critical role of automation technology and job contraction, balanced by the renewal of specific roles through upskilling and reskilling. The impact of automation on production efficiency and women's shifting roles was of particular interest, with demographics and geographic segmentation examined to determine local economic impacts. This pre-pandemic analysis heavily influenced strategic planning, formulation, and control mechanisms.

The onset of the COVID-19 era forced executives and managers to shift from strategic planning to immediate crisis management ('firefighting') in decision-making and implementation. This shift resulted in mixed outcomes; whilst swift and agile strategic decision-making was anticipated, often absent or reactive, leading to temporary business closures (Hyder, 2021). The future of work arrived rapidly in the US, establishing a 'new normal' with no reversion to pre-pandemic baselines.

The Future of Work in the Post-Pandemic Era

As the FOW evolves, US companies are accelerating digitalisation and automation, increasing the demand for advanced skills within a workforce that currently lacks adequate training. A significant skills gap is emerging: demand for physical and manual skills requiring repetition is projected to fall by approximately 30%, whilst demand for basic literacy and numeracy skills will decline by 20%. Conversely, demand for technological skills, including coding, could increase by approximately 50% (Ellingrud et al., 2020).

Flexible working arrangements, remote work, 'work-from-anywhere' policies, and employee benefits are increasingly shaping the US economy. Addressing critical business strategy elements such as profit margins, operations, customer engagement, and retention is equally vital. Failure to address these operational imperatives may compromise FOW transitions, as companies risk failure due to workforce skill shortages and unproductive operations. The US is currently operating under a strategy of radical uncertainty ('anything can happen'), necessitating high vigilance regarding macro-environmental shifts. Consequently, the US FOW approach focuses on employer-driven initiatives to ensure the right talent is in place to execute roles efficiently (Hyder, 2021).

Emotional Intelligence and Social Skills

Emotional intelligence and social skills are increasingly critical due to rising demand for initiative, critical thinking, leadership, entrepreneurial attitudes, empathy, and collaboration (Ellingrud et al., 2020). Consequently, the physical and mental well-being of the workforce has emerged as a priority. This shift has driven the adoption of remote and hybrid working environments, introduced by approximately 25–30% of companies, with some opting for fully remote models for the foreseeable future.

Concurrent with these developments, the four-day work week is gaining traction as employees demand improved working conditions. Companies such as Qwick, ThredUP, and US-based subsidiaries of Japanese firms like Panasonic and Toshiba have implemented this pattern. Since 2017, Basecamp has operated a four-day work week (during summer months) where employees work 32 hours without pay reduction (Carbonaro, 2022; Qwick, 2022). Increasingly, the four-day work week serves as a competitive benefit for recruitment and retention. Legislative efforts in California have even proposed standardising the 32-hour work week (Frias, 2023).

Despite these trends, some employers advocate for a full return to the office, arguing for increased productivity. However, this stance often triggers resignation decisions, as employees dispute that office presence equates to productivity, citing the benefits of eliminated commute times and reduced stress. Research indicates that strict return-to-office mandates are unpopular, with the hybrid model remaining the most favourable arrangement (Bloom et al., 2023).

The Great Resignation

Since the pandemic, the US has witnessed a significant increase in resignations, with 2.6% of the total workforce resigning by October 2022. The hospitality and leisure industry was most severely affected, experiencing a 2% quit rate (Ellerbeck, 2023). This phenomenon, termed 'The Great Resignation', describes the mass exodus of workers globally. In the US, high quit rates were concentrated in retail trade, healthcare, social assistance, and accommodation and food services, often driven by a loss of purpose and job fulfilment.

In 2021, the primary drivers (Figure 3.3) for quitting included low pay, disrespect from management, and career stagnation and factors associated with traditional work models. Pandemic-specific factors included re-prioritising personal life, childcare needs, health concerns, limited flexibility, and lack of well-being support (Ellerbeck, 2023). Consequently, the FOW requires a workplace reorientation that addresses these employee concerns. Short tenure (staying with an employer for less than two years) has become normalised as workers seek better salaries and conditions.

Figure 3.3: America's Top Quitting Factors

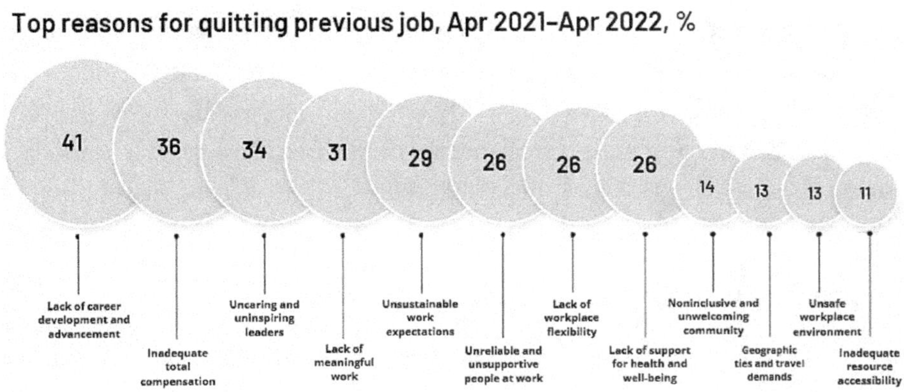

Source: Data from McKinsey (2022).

This Trend has been Described as a Fundamental Shift:

"It is the quitting trend that just won't quit. People are switching jobs and industries, moving from traditional to non-traditional roles, retiring early, or starting their own businesses. They are taking time out to tend to their personal lives or embarking on sabbaticals. The Great Attrition has become the Great Renegotiation." (Ellerbeck, 2023)

Impact of Rising Cost of Living

The rising cost of living, driven by inflation and interest rate increases, erodes earnings and purchasing power, drastically shifting employee perceptions of work. Workers increasingly value flexibility and 'freedom' over purely monetary gain, leading to growth in the gig economy. Employers must therefore work harder to convince talent to buy into their organisational vision, as skills become more transferable. Compensation and reward systems that recognise macro-environmental challenges are essential, given that 33% of workers lack trust in CEOs and 31% in line managers (Korn Ferry, 2023), and favourable reward programmes serve as a critical compensatory factor.

UNSDG 10: Reduced Inequality

The current US workplace learning system appears insufficiently inclusive. Minorities requiring the most training are often excluded from reskilling and upskilling programmes (Somers, 2022), exacerbating the wealth gap as the FOW unfolds. Newly created jobs requiring advanced skills face shortages, creating a training disparity that hinders recruitment, whilst those from advantaged backgrounds continue to progress.

The pandemic widened this divide, as remote work often deprioritised training. Frontline workers (e.g., security personnel, cleaners) are most affected; only 38% receive necessary training. Although the gig economy is surging (Ellerbeck, 2023), contract and freelance workers often lack access to upskilling programmes, leading to a long-term skills shortage. Government support for inclusive upskilling is therefore critical to ensuring a sustainable workforce.

America's Proposal for Workforce Revamp

The Trump administration has issued a comprehensive workforce development strategy aimed at transforming the US labour market. This strategy evaluates the needs of American companies and integrates existing systems for greater efficiency (Maurer, 2025). Five pillars have been identified as at the core of transforming America's workforce (see Figure 3.4).

Figure 3.4 America Workforce Vision

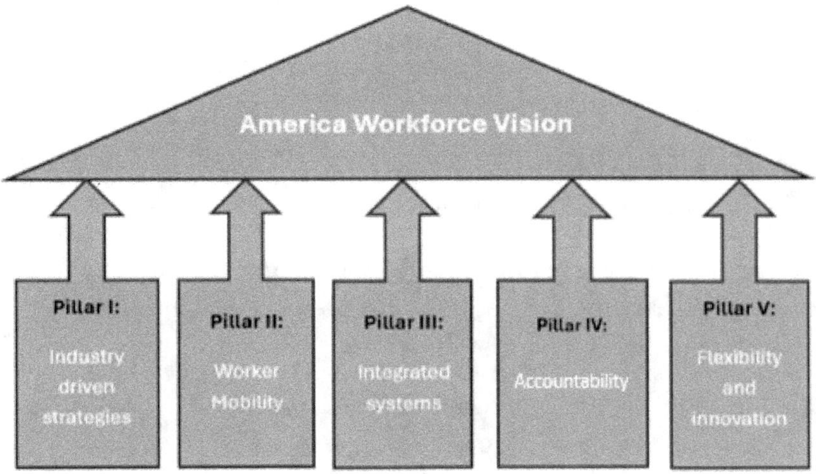

Source: Adapted from US Department of Labour (2025)

These five strategic pillars focus on increasing work-based models to connect people to in-demand jobs, streamlining federal workforce development programmes, and ensuring accountability in federally funded initiatives. The ultimate goal is to ensure the American workforce is prepared for an AI-driven economy.

Currently, training access remains uneven. Half of Hispanic workers and 32% of high school-educated workers receive only informal training. Overall, 55% of workers receive formal employer training, whilst 43% rely on informal methods. Trends indicate a shift towards community colleges and self-directed learning (e.g. YouTube) for reskilling. However, those with lower educational levels risk being left behind (Ellerbeck, 2023; McKinsey Global Institute, 2019). The FOW threatens to blur class distinctions, potentially pushing the lower-middle class towards poverty (Miami, 2023).

Automation and Digitalisation in the Future of Work

The COVID-19 pandemic accelerated the implementation of automation and digitalisation, leaving employers struggling to recruit workers with the requisite skill sets. This skills shortage is expected to persist. Automation and AI adoption threaten middle-class status by automating production and routine tasks, reducing remuneration and opportunities for those who do not reskill.

This shift may restrict access to quality education for the children of displaced workers, further widening the socio-economic gap (Miami, 2023).

The 'America Workforce Strategy' encourages employers to integrate systems that enhance worker skills to meet these demands (US Department of Labor, 2025).

Artificial Intelligence (AI) Barometer USA

Despite a weakening job market in 2024, the US recorded noticeable growth in AI-related skilled jobs. By 2022, the US posted the highest number of AI jobs globally (Figure 3.5).

Figure 3.5 US 2021–2024: Total Number and Share of Job Postings Requiring AI Skills

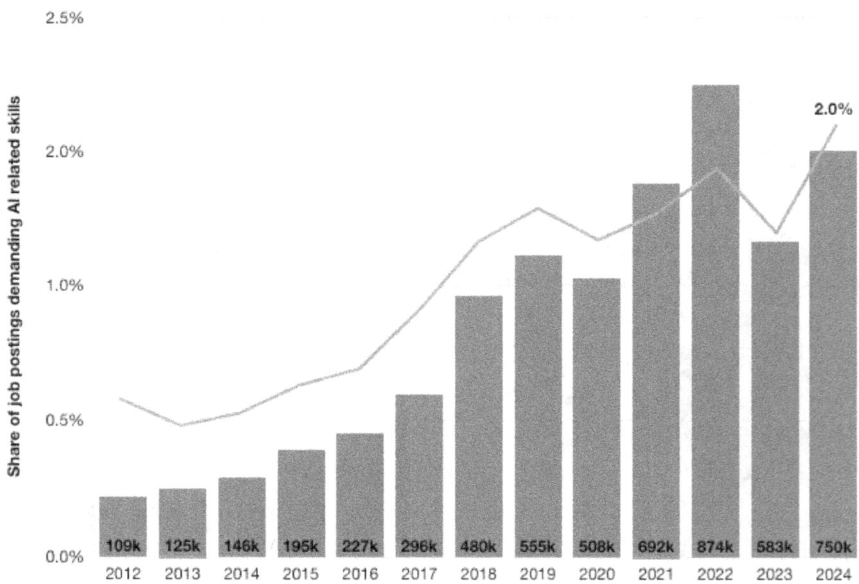

Source: Data from PwC (2025).

Between 2019 and 2024 (Figure 3.6), the US experienced a 20% increase in jobs less exposed to AI, such as health associate professionals, cleaners, and food preparation assistants. This trend mirrors patterns in the UK. Conversely, jobs associated with Generative AI (e.g., CEOs, senior officials, ICT professionals) recorded slower growth (1%) over the same period (PwC, 2025). However, demand for AI skills in the communication technology sector outpaces all other sectors, followed by professional and financial services.

Figure 3.6: US 2019–2024: Cumulative Growth Rate in All Job Postings Against Exposure to AI

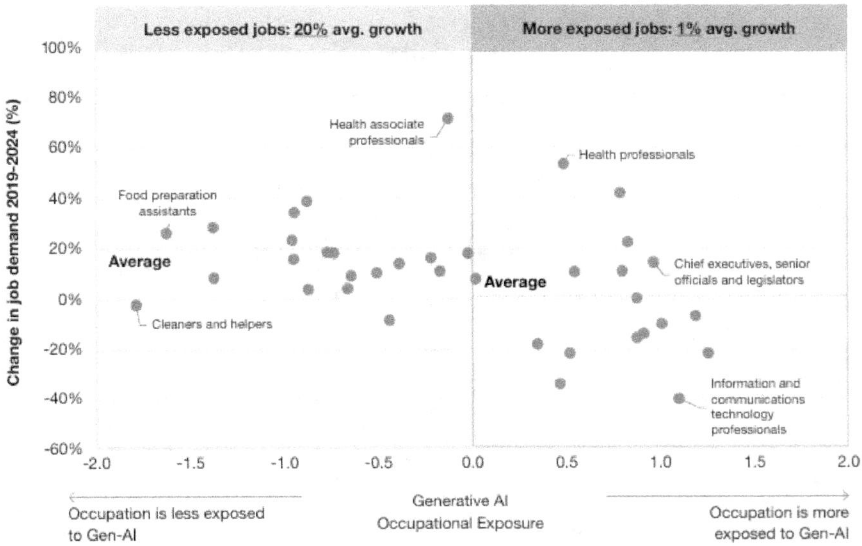

Source: Data from PwC (2025).

Despite slower overall growth in AI-specific roles, there is a noticeable increase in 'AI-augmented' jobs compared to 'AI-automated' jobs. This suggests that human expertise ('human-in-the-loop') remains valued as technology develops. Jobs in scientific and technical activities have experienced significant net skill changes (over 10 skills per posting), indicating rapid adaptation. In contrast, the education and health sectors show slower augmentation skill changes, reflecting lower AI adoption rates.

Notably, degree requirements for high AI-exposed jobs decreased from 63% in 2019 to 53% in 2024. Conversely, in lower AI-exposed jobs, there was a small decrease in degree requirements (2019-13%; 2024-12%). However, those within the top half of AI and augmentation exposure require a degree four times as often as those in the bottom half.

In AI-automated and augmented jobs, the requirement for a degree is drastically falling, although the majority of jobs in the US still list degree requirements, demonstrating a continuous reliance on formal education (PWC, 2025).

The Workforce

America Workforce Strategy also identified the shortcomings of the 'college-for-all' as not meeting the desirable skills and needs of both employer and employee. This is leading to a shortage of talent pipelines where most workers seem to be disconnected from high-wage career pathways. Some of the plans outlined in 'America Workforce Strategy: Building the workforce for the golden age' are to upskill the incumbent America and make them capable of keeping pace with AI developments and to develop alternatives to '4-year college degree' (Maurer, 2025). This confirms PWC barometer results that in the US, the requirement for a degree in the workplace is declining (PWC, 2025) as the nation prepares for a different future of work with different job entry requirements due to AI and the demand for work-based learning, where workers will continuously reskill and upskill.

FOW Employer Trends

FOW employer trends focus on training employees in workforce behaviour orientation and safety, but job training and progression are missing, which are at the forefront of employee job satisfaction. Manufacturing and biotech firms could train frontline workers to become prospective technicians, thereby preparing for the future, as 80% of HR executives increasingly find it difficult to recruit the right skilled personnel, a situation made worse by the lack of employer-provided training courses (Somers, 2022).

Should employers not provide training for their staff in a FOW with potential skills shortages, fierce competition for workers could arise, with those employers willing to train, reskill, and upskill workers able to attract the future worker. Large companies, with adequate resources, could offer such training or outsource it to keep employees prepared for career progression, as workers shun jobs lacking a clear progression path.

Recommendations

To address these challenges, the following strategic actions are recommended:

- **Inclusive Reskilling:** Reskilling and upskilling initiatives must be inclusive to ensure utility for all employees, regardless of grade, mitigating the risk of widening inequality.

- **Government Leadership:** The government must assume a stronger coordination role with stakeholders to ensure effective FOW implementation across the country.
- **Employer Responsibility:** Employers must recognise employee needs as paramount to attracting and retaining talent, integrating reskilling and upskilling into core organisational strategy.
- **Reframing Automation:** Automation should be embraced as a tool for augmentation and productivity improvement, rather than solely as a mechanism for job replacement.
- **Stakeholder Knowledge Transfer:** As automation and digitalisation create new roles, wider stakeholder engagement is required to transfer knowledge regarding FOW benefits.
- **Comprehensive Monitoring:** The wider impacts of the FOW require rigorous monitoring from top-level strategy down to operational implementation.

Chapter 4

Future of Work in the G7 – Canada

Introduction

Canada covers the majority of northern North America and is the second-largest country in the world by total area. Despite its vast size, approximately 80% of its landmass is uninhabitable due to climatic and geographic conditions. The country comprises 10 provinces and three territories, with Ontario and Quebec being the largest provinces, situated on the Canadian Shield. The capital city, Ottawa, is located in Ontario.

Figure 4.1: Toronto Skyline

Source: Getty Images

The majority of Canada's 40.5 million (Figure 4.2) population is concentrated near the southern border with the United States (US), Canada's only neighbouring country. Approximately 90% of Canadians live within 160 kilometres of the US border. The country's geography ranges from west to east and north to south, with the northern regions becoming progressively colder, rockier, and less hospitable (Canada Guide, n.d.; Worldometer, 2023). Canada is recognised as one of the world's wealthiest nations, possessing abundant natural resources.

Figure 4.2 shows Canada population trends with an overall increase in population for the period up to 2100.

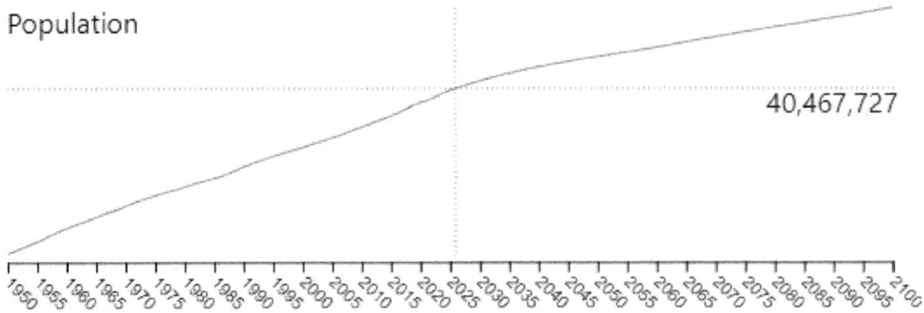

Source: Populationpyramid.net (2026)

The COVID-19 pandemic accelerated changes in the world of work across the globe, including Canada. The nature of work is evolving under the pressures of globalisation, macro-environmental factors, automation, digital transition, an ageing workforce, and Industry 4.0 technologies. Additionally, shifting work patterns such as work from home (WFH), work from anywhere (WFA), and flexible arrangements are redefining the Canadian employment landscape.

Table 4.1: shows Canada's Population trends for the period 1955 to 2026

Year	Population	Yearly % Change	Yearly Change	Migrants (net)	Median Age	Fertility Rate	Density (P/Km²)	Urban Pop %	Urban Population	Country's Share of World Pop	World Population	Canada Global Rank
2026	**40,467,728**	0.85%	341,005	291,403	40.8	1.33	4	80.2%	32,464,952	0.49%	8,300,678,395	38
2025	**40,126,723**	0.97%	384,293	326,204	40.6	1.33	4	80.2%	32,164,167	0.49%	8,231,613,070	38
2024	**39,742,430**	1.13%	443,325	368,599	40.5	1.34	4	80.2%	31,863,601	0.49%	8,161,972,572	38
2023	**39,299,105**	1.23%	477,846	433,842	40.3	1.35	4	80.3%	31,563,358	0.49%	8,091,734,930	38
2022	**38,821,259**	0.85%	324,679	459,988	40.3	1.37	4	80.5%	31,263,957	0.48%	8,021,407,192	38
2020	**38,171,902**	1.2%	441,934	256,592	39.9	1.41	4	80.3%	30,670,064	0.48%	7,887,001,292	38
2015	**35,962,234**	1.01%	353,067	224,332	39.6	1.6	4	81.2%	29,212,401	0.48%	7,470,491,872	38
2010	**34,196,899**	1.06%	351,345	225,369	38.7	1.64	4	80.9%	27,655,231	0.49%	7,021,732,148	36
2005	**32,440,173**	0.98%	309,674	204,526	37.6	1.57	4	79.7%	25,869,853	0.49%	6,586,970,132	35
2000	**30,891,803**	0.95%	286,534	201,727	35.8	1.51	3	79.1%	24,428,073	0.50%	6,171,702,993	34
1995	**29,459,131**	1.17%	333,938	141,818	33.7	1.67	3	77.3%	22,782,132	0.51%	5,758,878,982	33
1990	**27,789,443**	1.4%	372,396	181,461	31.8	1.7	3	76.3%	21,207,473	0.52%	5,327,803,110	32
1985	**25,927,465**	1.11%	278,738	69,776	29.9	1.59	3	76.2%	19,755,707	0.53%	4,868,943,465	32
1980	**24,533,776**	1.17%	276,958	122,743	28.1	1.66	3	75.7%	18,565,807	0.55%	4,447,606,236	32
1975	**23,148,986**	1.55%	341,793	139,797	26.4	1.8	3	75.6%	17,508,653	0.57%	4,070,735,277	31
1970	**21,440,022**	1.72%	350,901	304,627	25	2.27	2	75.7%	16,229,750	0.58%	3,694,683,794	30
1965	**19,685,518**	1.92%	357,346	112,401	24.5	3.08	2	73%	14,362,464	0.59%	3,334,533,703	29
1960	**17,898,790**	2.62%	434,103	49,694	25.5	3.82	2	69.1%	12,373,883	0.59%	3,015,470,894	30
1955	**15,728,274**	2.74%	397,305	95,995	26.3	3.75	2	65.7%	10,340,832	0.57%	2,740,213,792	30

Source: National Post (n.d.).

Canada's Population and the Future of Work

Canada's population is growing steadily (Table 4.1), providing a potential pool of future workers, provided they can be reskilled and upskilled to meet the demands of the future of work. As shown in Figure 4.3, the ratio of rural to urban population has shifted significantly between 1955 and 2023, with steady urbanisation and the majority of the population now residing in cities.

Figure 4.3 Canada Urban vs Rural Population (1955–2023)

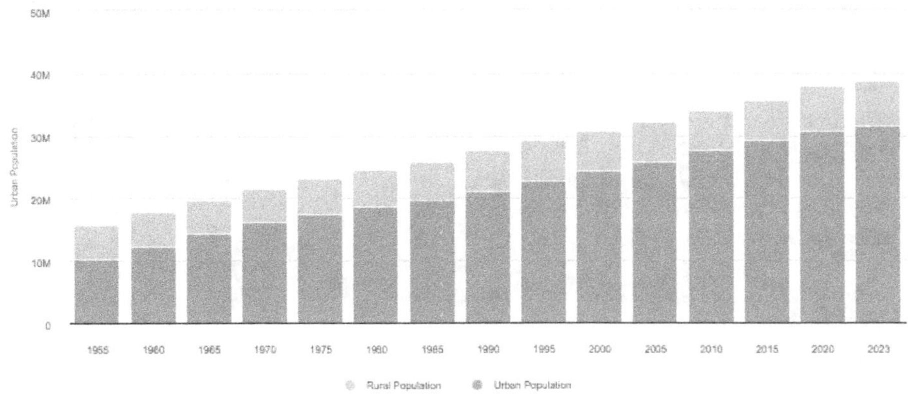

Source: Data from Worldometer (2023).

Table 4.2 illustrates Canada's annual population change from 1965 to 2023. Steady growth was recorded until 1999; however, the year 2000 marked a slight decrease. Growth fluctuated subsequently until the onset of the COVID-19 pandemic in 2020, which likely contributed to a further decline. Historical data highlights the significant role of migration in maintaining growth rates. The median age is rising, reaching 40.6 years in 2023.

Table 4.2: Canada Population: Current (2023) and Historical Trends

Year	Population	Yearly % Change	Yearly Change	Migrants (net)	Median Age	Fertility Rate	Density (P/Km²)	Urban Pop %	Urban Population	Country's Share of World Pop	World Population	Canada Global Rank
2023	**38,781,291**	0.85 %	326,964	249,746	40.6	1.47	4	81.4 %	31,563,358	0.48 %	8,045,311,447	38
2022	**38,454,327**	0.78 %	299,315	248,586	40.4	1.47	4	81.3 %	31,263,957	0.48 %	7,975,105,156	39
2020	**37,888,705**	0.98 %	366,121	195,181	39.9	1.46	4	80.9 %	30,670,064	0.48 %	7,840,952,880	39
2015	**35,732,126**	1.02 %	353,743	232,738	39.6	1.55	4	81.8 %	29,212,401	0.48 %	7,426,597,537	38
2010	**33,963,412**	1.06 %	349,499	227,724	38.7	1.63	4	81.4 %	27,655,231	0.49 %	6,985,603,105	35
2005	**32,215,916**	0.98 %	306,521	204,710	37.6	1.54	4	80.3 %	25,869,853	0.49 %	6,558,176,119	35
2000	**30,683,313**	0.93 %	278,778	201,617	35.8	1.49	3	79.6 %	24,428,073	0.50 %	6,148,898,975	35
1995	**29,289,423**	1.15 %	326,444	137,472	33.8	1.68	3	77.8 %	22,782,132	0.51 %	5,743,219,454	33
1990	**27,657,204**	1.37 %	362,656	145,270	31.9	1.81	3	76.7 %	21,207,473	0.52 %	5,316,175,862	32
1985	**25,843,922**	1.06 %	266,482	39,796	30.0	1.67	3	76.4 %	19,755,707	0.53 %	4,861,730,613	32
1980	**24,511,510**	1.16 %	275,052	101,270	28.1	1.74	3	75.7 %	18,565,807	0.55 %	4,444,007,706	32
1975	**23,136,252**	1.54 %	340,335	130,169	26.4	1.85	3	75.7 %	17,508,653	0.57 %	4,069,437,231	31
1970	**21,434,577**	1.71 %	349,154	295,787	25.0	2.33	2	75.7 %	16,229,750	0.58 %	3,695,390,336	30
1965	**19,688,805**	1.91 %	355,890	103,588	24.5	3.14	2	72.9 %	14,362,464	0.59 %	3,337,111,983	30

Source: Adapted from Worldometer (2023).

The Canadian workforce is ageing rapidly, posing challenges for future job skills requirements. From 2026, about 4 in 10 working Canadians will be aged 55 or older (Fields et al., 2017). This demographic shift will impact labour market participation rates and place increased pressure on government and healthcare systems.

Figure 4.4: Canada Annual Population Growth Rate (1950–2020)

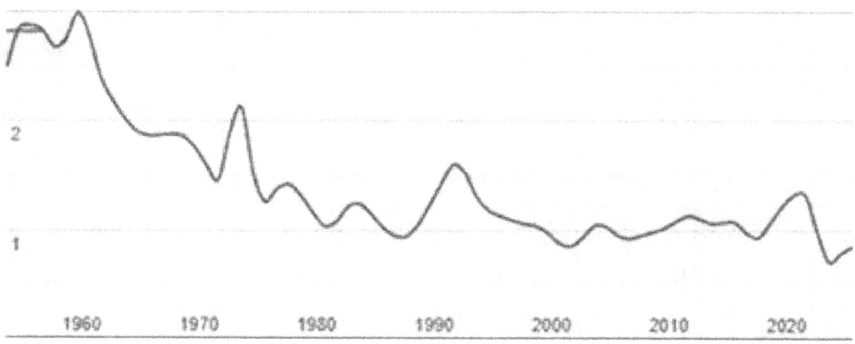

Source: Data from Worldometer (2023).

Although Canada's population growth rate is slowing (see Table 4.3), forecasts indicate a marginal increase through 2050. This contrasts with other G7 nations, such as Italy, which faces negative population growth projections. A positive growth rate is critical for facilitating skills planning and integrating new workforce entrants into emerging industry sectors.

Table 4.3: Canada Population Forecast (2025–2050)

Year	Population	Yearly % Change	Yearly Change	Migrants (net)	Median Age	Fertility Rate	Density (P/Km²)	Urban Pop %	Urban Population	Country's Share of World Pop	World Population	Canada Global Rank
2025	**39,431,447**	0.80 %	308,548	249,808	41.0	1.48	4	81.6 %	32,164,167	0.48 %	8,191,988,453	38
2030	**41,008,596**	0.79 %	315,430	258,947	42.0	1.50	5	82.1 %	33,662,820	0.48 %	8,546,141,327	38
2035	**42,503,204**	0.72 %	298,922	259,193	43.0	1.51	5	82.6 %	35,120,468	0.48 %	8,879,397,401	41
2040	**43,822,969**	0.61 %	263,953	259,379	43.9	1.53	5	83.4 %	36,540,219	0.48 %	9,188,250,492	43
2045	**44,932,177**	0.50 %	221,842	249,998	44.8	1.54	5	84.4 %	37,921,907	0.47 %	9,467,543,575	46
2050	**45,890,819**	0.42 %	191,728		45.3	1.55	5	85.5 %	39,233,850	0.47 %	9,709,491,761	45

Source: Adapted from Worldometer (2023).

Technological Digitalisation, Artificial Intelligence, and Demographics

Emerging technologies such as Virtual Reality (VR), Artificial Intelligence (AI), and robotics are disrupting workforce composition. VR is becoming a workplace reality, offering cost-effectiveness and diverse applications. These technologies will blur the boundaries between living and working spaces, potentially merging national service sectors with global ones. This dissolution of boundaries creates complexities for tax and labour laws, whilst reduced travel demand will impact urban design and city planning.

AI-based decision-making is gradually replacing traditional knowledge-based roles in both the public and private sectors. This shift may accelerate 'de-credentialisation', the declining economic value of traditional academic credentials (Amery & Bonen, 2018). The rise of the gig economy is replacing long-term employment arrangements with flexible, multiple-income stream models. Although automation will result in job displacement, most occupations will undergo restructuring rather than total elimination. Specific tasks will be automated, requiring staff to upskill (Guliyev et al, 2023).

McKinsey & Company (2017) reported that 9% of total employment is at risk of full automation. Lamb and Doyle (2017) estimated that 42% of the Canadian workforce will be affected by automation over the next 10 to 20 years, with low-skilled workers more susceptible than highly qualified professionals. Amery and Bonen (2018) further noted that technological advancements will impact approximately 10 % of the workforce.

The labour market will see greater participation from senior citizens, who may use the gig economy to supplement retirement income (Fields et al., 2017). Corporations must revise human resource (HR) policies to adopt age-inclusive hiring. In this digital era, an individual's digital reputation, measurable via AI, will become a tangible asset and a differentiator in the labour market. Demand will increase for candidates possessing strong critical thinking and analytical skills (Lillico, 2017).

Artificial Intelligence Barometer: Canada

Canada has experienced strong growth in AI-related job postings, rising from 4,000 in 2012 to 57,000 in 2024. The most significant increases occurred in 2023 (33,000) and 2024 (57,000) (PwC, 2025). Canada is unique among G7 nations in experiencing a sharp increase in AI-skills postings within the Financial and Insurance sector, followed by Information and Communication. Education ranks third, followed by Professional, Scientific, and Technical activities (see Figure 4.5).

Figure 4.5: Share of AI Job Postings by Sector in Canada (2012–2024)

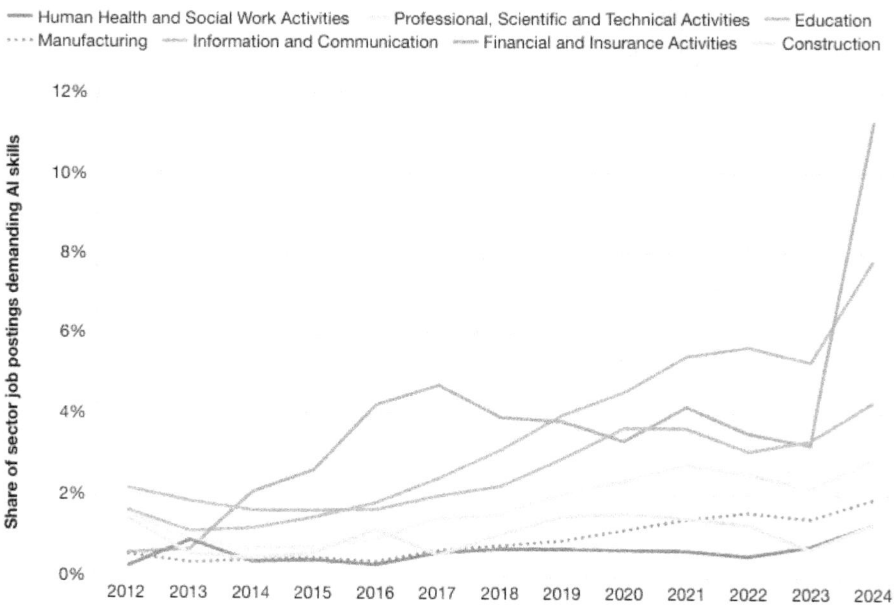

Source: Data from PwC (2025).

There appears to be no direct correlation between AI occupational exposure and growth in job postings (2019–2024), suggesting AI exposure has had a limited systematic effect on demand. Both low- and high-exposure jobs show varied growth rates, implying other factors drive demand (see Figure 4.6). Notably, jobs with higher AI exposure have grown at a slower rate than those with lower exposure (PwC, 2025).

Figure 4.6: Cumulative Growth Rate in Job Postings vs AI Exposure (2019–2024)

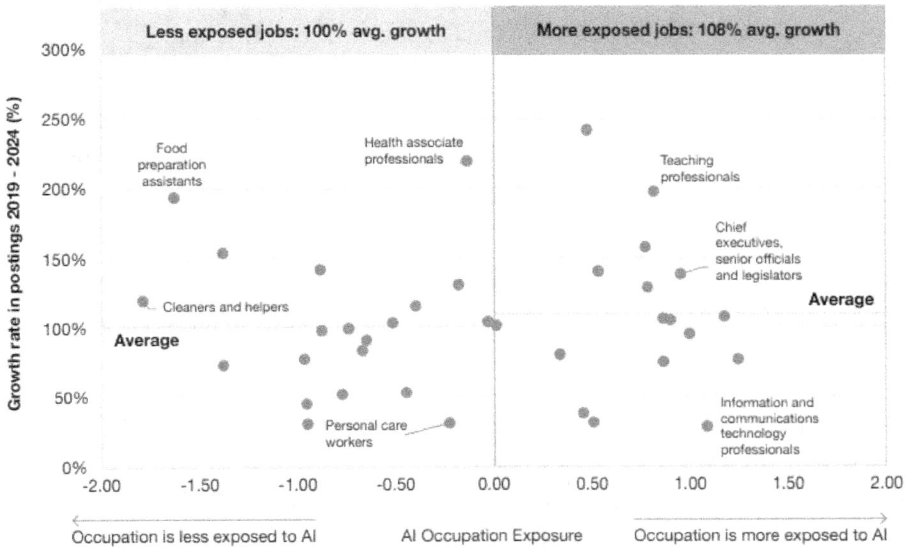

Source: Data from PwC (2025).

Canada's Health and Social Work sector leads in AI job growth, with augmentation rates at 180% and automation above 130%, significantly surpassing national averages (87% for augmentation and 65% for automation). This indicates a strong AI transformation in healthcare. However, slower growth is evident in Finance, Information and Communication, and Mining. Agriculture records the slowest growth rates. This disparity suggests that, outside of healthcare, Canada's pace of AI transformation and workforce reskilling may be lagging compared to other G7 nations.

Interestingly, degree requirements for jobs with high AI exposure have declined by 13%, falling from 70% in 2019 to 57% in 2024. A similar trend is observed in augmentation-exposed jobs (75% to 65%) and low-exposure jobs (14% to 11%). Whilst degree requirements are falling, they remain significantly higher for the top 50% of most exposed jobs (PwC, 2025).

Employment, Income Distribution, and Climate Change

Climate change will fundamentally impact employment, growth, and income distribution. Amery and Bonen (2018) estimated that failing to mitigate climate change could cost Canada CAD 21–43 billion annually by 2050. The transition to a green economy involves phasing out polluting industries and expanding green sectors (Barbier, 2016). This structural change will create substantial job opportunities but will also disrupt the Canadian labour market. .

Future of Work Patterns

The Canadian future of work scenario is evolving, with a majority of Canadians favouring WFH. A 2022 poll of 6,600 individuals indicated a slightly stronger preference for WFH among males (77%) compared to females (23%) (Subramaniam, 2022). Among 1,600 surveyed employees, only 12% favoured a full return to the office, whilst 48% stated they would leave their job if forced to return (Amazon Canada, 2022), meaning 40% prefer a hybrid work model.

Employees report better work-life balance when working from home, believing pre-COVID structured patterns do not align with modern life. This shift renders some physical office space redundant (Chan, 2021; Subramaniam, 2022). The concept of 'work-from-anywhere' is gaining traction, facilitated by VR and technology that create global workspaces. This allows employers to access a global talent pool, potentially merging global wages but challenging existing tax frameworks.

Table 4.4: Current Work Location by Selected Characteristics (Aged 15–64) in Canada (2021)

	Working from home all of the time	Working from home some of the time	Working at workplace outside home all of the time
		%	
Total	**20.6**	**17.8**	**61.6**
Gender			
Men	19.1	16.6	64.3
Women	22.6	19.3	58.1
Age group			
15 to 34 years	19.1	15.5	65.4
35 to 54 years	22.0	19.1	58.9
55 to 64 years	19.6	18.4	62.0
Rural vs. urban			
Rural	13.5	11.3	75.2
Urban	21.6	18.7	59.7
Region			
Atlantic provinces	15.1	16.4	68.5
Quebec	21.1	18.3	60.6
Ontario	25.7	17.8	56.5
Prairie provinces	14.8	17.3	67.9
British Columbia	16.2^E	18.2^E	65.6
Highest level of education			
High school diploma or below[1]	9.5	8.1	82.4
Postsecondary certificate or diploma below the bachelor's level[2]	17.1	17.8	65.1
University certificate, diploma or degree at bachelor's level or higher	32.5	25.0	42.5

Source: **Adapted from Statistics Canada (2021). High school diploma includes equivalency. Postsecondary certificate includes trade certificates and college diplomas below bachelor's level.**

Job Redesign and Realignment

Jobs are being recategorised into eight groups: knowledge workers, personal services, supervisors, technical trades, non-technical trades, builders, and doers, effectively dissolving the traditional white-collar versus blue-collar distinction (Conference Board of Canada, 2020a). Technological advancements are driving structural changes across industries, suggesting that a return to pre-COVID norms is unlikely.

Figure 4.7: Canada's Hardest Hit Occupations in the First Year of COVID-19 (2019–2020)

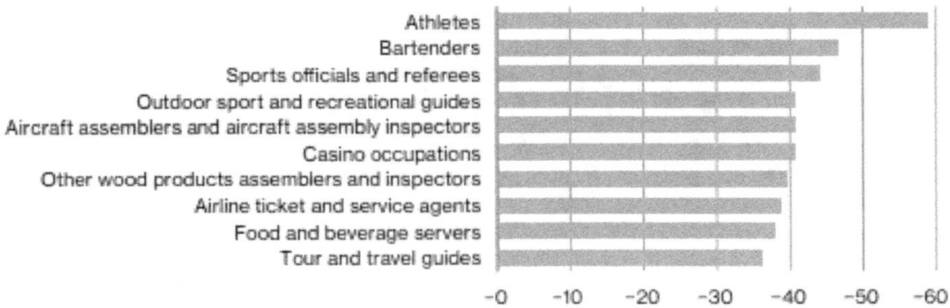

Source: Data from The Conference Board of Canada (2022).

In the wake of the global socio-economic shock brought about by COVID-19, the Canada FOW model across various industries indicates that the health and well-being of both employees and customers are paramount (Canada Beyond, n.d.). Figure 4.7 shows the hardest hit jobs during the initial phases of the pandemic (2019-2020).

Recommendations

- **Lifelong Learning:** Implement a culture of lifelong learning across schools, universities, and corporations. Develop flexible training programmes and collaborative labs to match curricula with evolving skills requirements.
- **Policy Reform:** The government should evaluate policy changes regarding retirement age to enable continued labour market participation by older citizens.
- **Digital Regulation:** Develop regulations to address challenges related to digital footprints, surveillance, privacy, and censorship.

- **Career Mobility:** Move away from the 'career ladder' model towards a 'lattice' model. Develop competency frameworks that facilitate lateral work-related transitions.
- **Flexible Work:** Adapt HR roles to manage the demand for 'work from anywhere' and flexible arrangements.
- **Job Re-categorisation:** Formally recognise the blurring of white-collar and blue-collar classifications in job design.
- **Technological Integration:** Utilise technology to inform job redesign and hiring processes.

Chapter 5

Future of Work in the G7 – The United Kingdom

Introduction

The United Kingdom (UK) is an island nation located off the northwestern coast of mainland Europe. It comprises four countries: England, Scotland, Wales, and Northern Ireland. Distinctively, Great Britain refers to the landmass comprising England, Scotland, and Wales, excluding Northern Ireland. The UK is renowned for its parliamentary democracy and historical role as the birthplace of the Industrial Revolution. London, its capital, serves as a world-leading commercial, financial, and cultural centre, celebrated for its rich history, contemporary arts, and iconic monuments such as the Houses of Parliament and Big Ben (Whitlock et al., 2024).

Figure 5.1: Big Ben and the Houses of Parliament

Source: Getty Images.

UK Population Trends

The UK population has demonstrated consistent growth since 1982 (see Figure 5.2), a positive indicator for future workforce talent availability, in contrast to the declining demographics observed in other G7 nations such as Italy. Projections estimate the population will rise to 73.7 million by 2036 and 76.6 million by mid-2046 (Office for National Statistics [ONS], 2023).

Historically, natural change drove population trends. However, since the 1990s, net migration has become the primary driver of growth. The impact of the COVID-19 pandemic caused significant demographic disruption: in 2020, the number of births fell to its lowest level since 2002 (decreasing by approximately 31,120 from 2019), whilst the number of deaths reached 689,629, the highest figure recorded since 1918 (Robards, 2024).

Figure 5.2: UK Population Estimates (Mid-1996 to Mid-2021) and Projections to Mid-2046

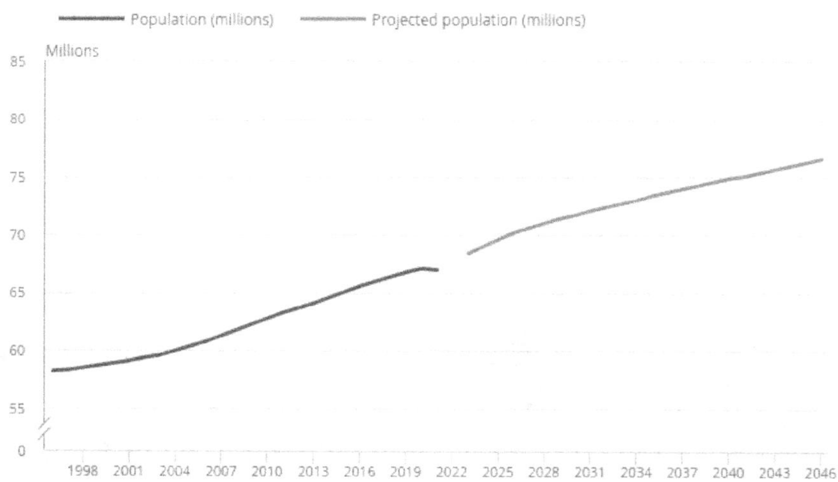

Source: Data from Office for National Statistics (2023).

International migration patterns have also shifted. Figure 5.3 illustrates long-term international migration trends up to 2020. Critical events, notably the European Union (EU) referendum (Brexit) and the COVID-19 pandemic, significantly impacted net migration (Gilbert-Johns et al., 2022). The pandemic precipitated a decline across immigration, emigration, and net migration metrics. Furthermore, the departure of EU citizens during this period exacerbated skills shortages in specific sectors, particularly hospitality.

Figure 5.3: Long-Term Immigration, Emigration, and Net Migration

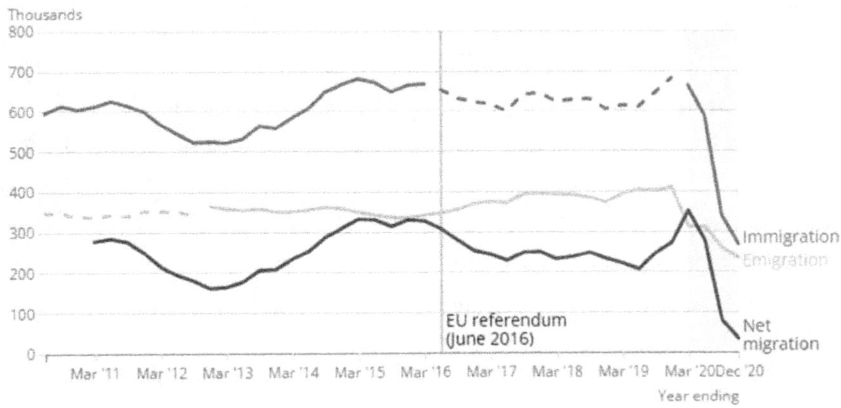

Source: Data from Office for National Statistics (2022).

The Dual Impact of COVID-19 and Brexit

The convergence of COVID-19 and Brexit triggered drastic socio-economic changes across the UK. The pandemic resulted in significant loss of life and necessitated lockdowns that disproportionately affected low-income earners as the economy contracted (Arrieta, 2022; Rao Madhu Sudhana et al., 2022; Wolf, 2022). Simultaneously, Brexit altered the economic landscape, complicating recruitment and trade relationships with EU member states. This dual shock accelerated economic recession at an unprecedented scale (Baryshnikova et al., 2021; Harari et al., 2021; Kraemer et al., 2020). Many organisations responded by reducing workforce numbers to survive, aggravating employment instability (Kolluru et al., 2021).

These crises hastened digital transformation. Companies frantically embraced digitalisation to adapt, compressing decades of anticipated technological adoption into a short period (PwC, 2022). Terms such as 'intelligent digital', 'artificial intelligence' (AI), and 'augmented reality' (AR) became central to defining the transition into the future of work, signalling an irreversible shift in operational models (Cameron, 2022; Gtmhub, 2023).

Digital Disruption, Innovation, and Agility

The post-pandemic UK workforce is smaller than pre-2020 levels, diverging from trends in France, Germany, and Spain, where workforce numbers have rebounded (Hill, 2022). The future of work in the UK is characterised by increasing integration of AI, robotics, and 'smart' automation. These technologies promise productivity gains and the creation of new products and services.

Innovation is fundamental to this transition. Employees require autonomy and access to technology to innovate, necessitating a cultural shift where employers tolerate failure as part of the learning curve. This approach facilitates cost-effective innovation and production efficiency. Furthermore, organisations must develop core competencies in sustainable production to reduce carbon footprints, aligning with United Nations Sustainable Development Goal (UNSDG) 12: Responsible Consumption and Production (Eckhardt et al, 2019).

In the short term, organisations may experience reduced profit margins (Coydel, 2022) and job displacement due to digital disruption. The narrative of 'are robots stealing our jobs' reflects genuine concerns regarding automation's mixed impacts (PwC, 2018; Rao Madhu Sudhana et al., 2022).

Three Waves of Automation

The evolution of automation in the UK is categorised into three distinct waves (Table 5.1):

1. Algorithmic - Wave 1 (Early 2020s): Algorithmic. Automation of simple computational tasks and structured data analysis, prevalent in data-driven sectors like financial services.

2. Augmentation - Wave 2 (Late 2020s): Augmentation. Current phase involves deeper human-technology interaction for clerical support and decision-making. Robotics is increasingly used in semi-controlled environments (e.g., warehouses, aerial drones).

3. Autonomy - Wave 3 (Mid-2030s): Autonomy. Significant reduction in manual labour and physical dexterity requirements.

Table 5.1: UK Job Automation Rates by Industry, Gender, and Education Level (Waves 1–3)

	Wave 1 (to early 2020s)	Wave 2 (to late 2020s)			Wave 3 (to mid-2030s)				Total 3 Waves
	: algorithmic	: augmentation			: autonomy				
	Job automation potential	Automation potential by type of worker			Automation potential by type of worker				Algorithmic / Augmentation / Autonomy
Type of sector / industry	%	Male %	Female (%)	Total	Higher Education	Medium Education	Low Education	Total	Grand Total
All Industry sectors	2	1	4	5	3	2	3	8	15
Manufacturing	2	1	6	7	2	3	0	5	14
Construction	2	1	18	19	0	1	5	6	27
Wholesale and retail trade	2	1	3	4	1	2	3	6	12
Education	1	0	1	1	1	0	0	1	3
Human health and social work	1	1	1	2	0	1	5	6	9
Total	**10**	**4**	**29**	**33**	**4**	**7**	**13**	**24**	**65**

Source: Adapted from PwC (2018).

Global estimates suggest AI could contribute USD 15 trillion to the global economy by 2030. However, by the mid-2030s, approximately 30% of UK jobs and 44% of roles held by lower-educated employees are at risk of automation. Continuous reskilling is therefore essential to prevent workforce obsolescence (Rao Madhu Sudhana et al., 2022).

The UK's Wave 1 automation rate across all sectors was 2%. As the UK enters Wave 2 augmentation in the late 2020s, the impact will shift toward gender-based disparities, with 4% of female jobs projected for automation compared to only 1% of male jobs. The last wave of automation (mid-2030) will negatively impact 3% of jobs (HE graduates), 2% (medium education/College/Highers/A level), and 3% (low education, GCSE or lower) (PWC, 2018). In general, women's jobs are more susceptible to automation in the next decade, but men's jobs are at greater risk in the long run. The UK will experience varying degrees of automation across industry sectors. Financial services face high short-term risk due to the efficiency of algorithmic processing, whereas the transportation industry will experience automation over a much longer period. Generally, jobs with lower levels of education are the most at risk from automation and robotics. But the UK will face an ageing population that will benefit from mentoring by younger employees (PWC, 2022). To fully benefit from AI and robotics, highly educated workers will still be required; thus, STEM becomes critical to facilitate the absorption of new AI and robotics technological knowledge. Hence, the need for culture change to adapt to new jobs and careers requires reskilling and retraining. Soft skills will also be crucial in FOW to enable transitions and employability in new jobs. Table 5.1 details the projected automation rates. Wave 1 affected 2% of jobs across all sectors. Wave 2 (Augmentation) is projected to impact 1% of male and 4% of female jobs. Wave 3 (Autonomy) will impact 3% of graduates, 2% of medium-education workers, and significantly higher proportions of low-education workers (PwC, 2018).

Sectoral impacts vary. Financial services face high short-term risk due to algorithmic efficiency, whilst transportation automation will occur over a longer horizon. The construction industry faces the highest total automation risk (27%), followed by manufacturing (14%) and wholesale/retail (12%). Task analysis indicates that routine tasks (30% risk) and management (24% risk) are most susceptible to machine replacement.

To mitigate these risks, the UK must leverage its ageing workforce for mentoring whilst promoting STEM education to equip the younger generation with the necessary technical skills. Soft skills remain critical for employability in a technology-driven market (PwC, 2022).

Case Study: Robotics in Healthcare

Technological advancements in the health sector are exemplified by CMR Surgical, the UK's largest med-tech company, which deploys robotic surgery innovations. Robotic surgery is minimally invasive, reducing patient pain and recovery time (Kuchler, 2023). Increased adoption enhances operational efficiency, helping to address surgeon shortages. However, successful integration requires collaborative investment from government and organisations to provide appropriate training for medical professionals (Gartner, 2023; PwC, 2018).

Artificial Intelligence Barometer: UK

According to PwC's (2025) barometer, the UK is experiencing a rapid AI revolution (see Figure 5.4). Growth is evident even in traditionally manual sectors such as agriculture and construction. AI is democratising access to employment. Employers increasingly value AI exposure over formal degrees for certain roles. Between 2012 and 2024, demand for AI skills increased significantly across most sectors.

Figure 5.4 UK Share of AI Job Postings by Sector (2012–2024)

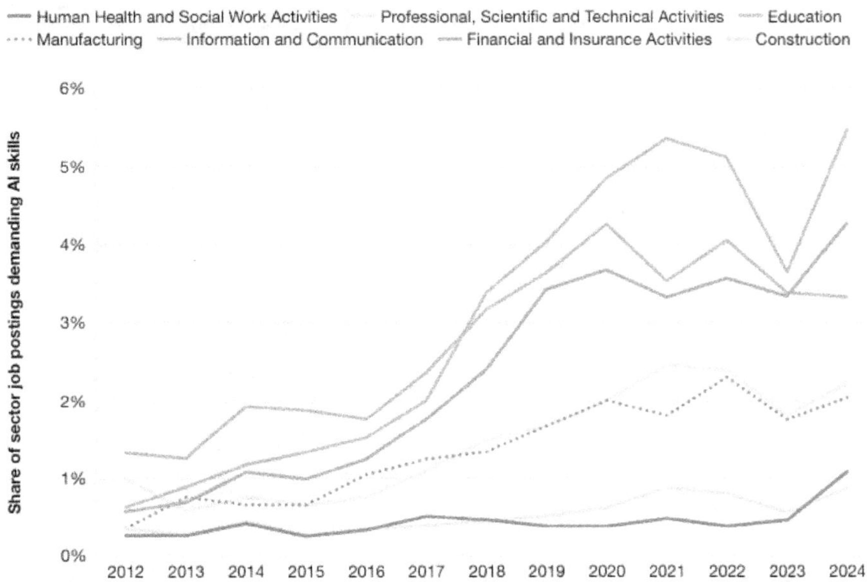

Source: Data from PwC (2025).

Interestingly, jobs less exposed to Generative AI (e.g., food preparation, cleaning, electrical trades) have seen higher average growth (47%) compared to highly exposed roles (17%), such as teaching and ICT professionals (see Figure 5.5).

Figure 5.5 Cumulative Growth Rate in Job Postings vs Generative AI Exposure (2019–2024)

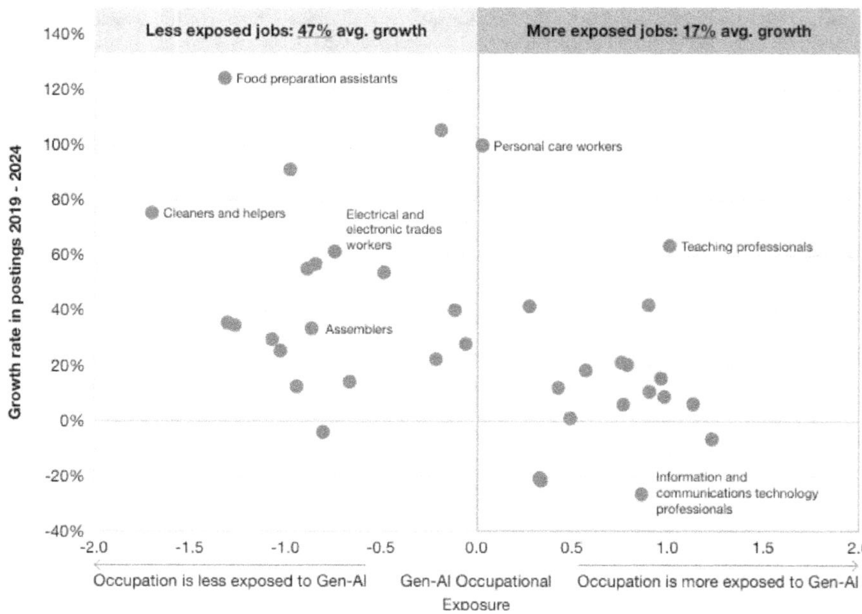

Source: Data from PwC (2025).

Degree requirements for high-AI-exposure jobs declined from 64% in 2019 to 56% in 2024. Lower AI exposure jobs experienced a smaller decline in degree requirements 1% between 2019 and 2024. This trend suggests a shift towards skills-based hiring, although the most exposed roles still require degrees at rates four times higher than low-exposure roles.

Organisational Agility

Future organisations in the UK will require agility across three levels:

1. **Organisational Agility:** Business models must align with strategy, supported by robust governance and data-driven decision-making will be expedited as employees are granted autonomy, thereby reducing the workforce. Access to big data will make decision-making more agile and data-driven rather than subjective (Rao Madhu Sudhana, Kolluru, and Hyams-Ssekasi, 2022).

2. **Workforce Agility:** Talent retention will depend on diverse incentives (flexibility, innovation, value-alignment). Agile operating models will likely

feature lean permanent structures supplemented by a transient, gig-based workforce (Hill, 2022; Wolf, 2022).

3. **Employee Agility:** The psychological contract will shift; employees will expect employers to maximise their talent for future progression. Professional reputation, based on project ratings, may replace the traditional CV (PwC, 2018).

Work Patterns, Health, and Well-being

Talent retention is a priority for UK employers. Many are adopting reduced working hours (e.g., a compressed four-day week) without pay reduction to meet employee demands for work-life balance (Hill, 2022; The Guardian, 2022). For example, Deloitte implemented a 'work-from-home' policy for its 20,000 UK employees, prioritising trust and flexibility (Deloitte, 2021). Globally, and particularly in the UK, resentment towards work will grow, and a significant number of employees will quit through the 'silent quitting' phase. However, rising living costs may force some workers to opt for reduced hours to save on childcare, prioritising family quality of life over career advancement. Whilst unemployment is forecast to remain low through 2026 (see Figure 5.6), automation poses a long-term risk unless effective reskilling initiatives are implemented (World Economic Forum, 2023).

Figure 5.6: UK Unemployment Rate Forecast

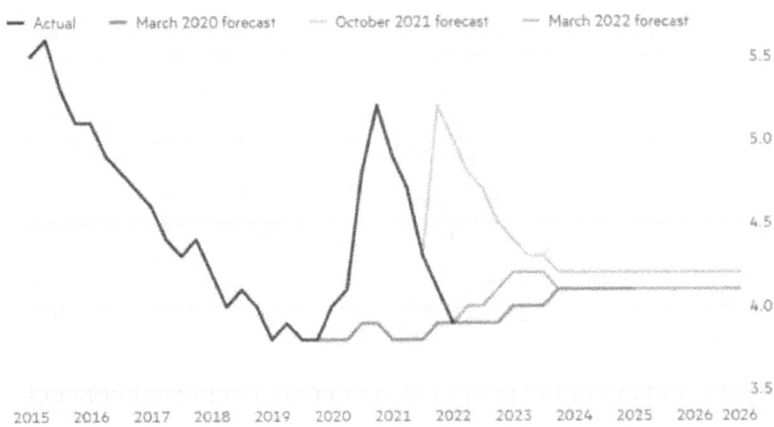

Source: Data from Office for Budget Responsibility, cited in Financial Times (2022).

Future Work Trends and Economic Impact

From 2024 onwards, work trends are redefining the employment landscape. 'In-person' requirements no longer strictly equate to office presence; 'work from anywhere' (WFA) allows teams to meet in diverse locations, repurposing office real estate. 'Workism', the defining of identity through employment, is declining as employees prioritise health and caregiving. Hence, employers who support caregiving and offer it as part of the job, as evidenced by their policies, become more attractive to special talent. The pursuit of the 'portfolio life', meaning putting work-life balance and wellbeing at the forefront, becomes the norm. Work, family, community, health, and rest become critical. Because of all these transformations, baby boomers are anticipated to pave the way for Gen X and millennials to become leaders in companies and governments. Unemployment will generally remain low (Figure 5.6). The injection of fresh talent, leadership styles, and thoughts on how these organisations and governments could be run could transform the future of work. But creativity and analytical thinking are super skills for future work, according to the Future of Jobs Report (World Economic Forum, 2025).

Generational Workforce, Nature of Work, Impact of Social Media, and GDP Effect

The highly skilled will opt for a better work-life balance, as monetary value alone would not be enough. Women become stronger and more represented in the workforce. Increasingly, due to being in Wave 3 (Table 5.1), more jobs become insecure as companies resort to contracts and outsourcing, and with a reduced permanent workers register (gig economy). UK employers believe the younger generation's turning up to work will be driven by the expectation of incentives to do work, as the pressure of Social Media will constantly promote the 'doing very little' motto through the drive called 'minimum wage and minimum effort, although wages might not be at minimum levels, resulting in reduced productivity. Social Media will become a dominant force for influencing and encouraging the 'doing less work' (Hill, 2022). Thus, using drones in some industry sectors, i.e., construction, due to the 'doing very little' motto, will result in £1.6 billion cost savings as productivity and efficiency improve, as workers no longer need to physically visit sites.

Using drones creates three GDP scenarios. Firstly, three outcomes are envisaged: a) output and GDP will increase as well as employees' pay resulting from upskilling in using drone technology information; b) output, profits, and GDP will increase; and c) output does not increase, demand is stagnant, production costs decrease, so does the wage bill, and GDP. Secondly, cost savings will lead to lower prices for products and services, benefiting consumers, and companies will gain from increased profits. Depending on the sector, some wage bills will decrease due to a reduced workforce as automation takes over (Hill, 2022; PWC, 2018). Lastly, some sectors' resources will face intense competition (investments and workforce), and some, due to increased profits, will pay higher wages to attract and retain talent, resulting in a decrease in GDP. Postal and Courier services will result in greater cost savings, but not in corresponding GDP growth, as cost savings are passed on to consumers (PWC, 2018). The Upstream Oil and Gas industry will experience little increase in GDP, and most of its workforce will transfer to other industries (see figure 5.7). Job insecurity and income will be the norm for the majority of workers; hence, the critical role of reskilling and upskilling as the world becomes more complex and more uncertain.

Figure 5.7 GDP and Cost Savings as a Percentage of 2030 GDP

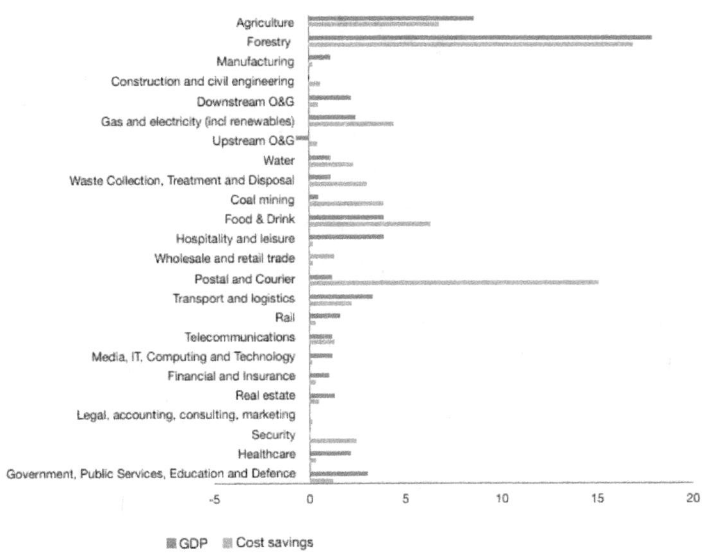

Source: Data from PwC (2018).

To guide the future of work, the UK government has identified 13 trends shaping the landscape to 2030 (see Figure 5.8), encompassing resources, law, society, and technology. These align with UNSDGs 3 (Health), 4 (Education), 8 (Decent Work), 12 (Consumption), and 13 (Climate Action).

Figure 5.8 Trends Shaping UK Jobs and Skills to 2030

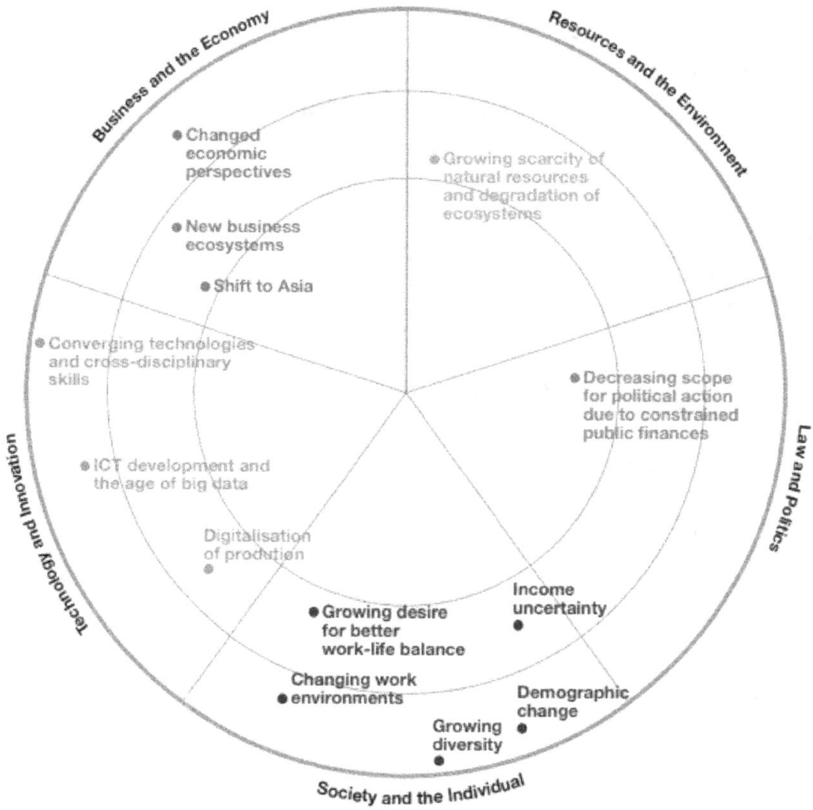

Source: Adapted from Government Office for Science (2017).

Hybrid working supports UNSDG-13 Climate by reducing commuting emissions (Okta, n.d.). However, recent mandates by companies such as THG plc returning employees to the office full-time threaten these environmental and flexibility gains (Wright, 2023). It remains to be seen how employees will react to the reversal of hybrid work imposed without notice.

Recommendations

To navigate these transitions effectively:

- **Collaboration:** Government and organisations must collaborate to facilitate workforce transitions through targeted reskilling.
- **Education Alignment:** Education providers must update curricula to match the economic demands of the future of work.
- **Social Safety Nets:** Government schemes are required to assist those displaced by automation, preventing poverty.
- **Human-Centric Approach:** Sectors reliant on human interaction (health, education) should prioritise well-being alongside automation.
- **Youth Incentives:** New incentives are needed to promote work ethics among the younger generation.
- **Economic Monitoring:** GDP growth and productivity must be closely monitored to assess the real impact of new work models.
- **Sustainability:** The integration of UNSDGs into business strategy is essential for long-term viability.

Chapter 6

Future of Work in the G7 – Germany

Introduction

Germany, the most populous nation in the European Union (EU) with approximately 84.5 million inhabitants, covers an area of 352,022 square kilometres. Unique among EU member states, it shares borders with nine countries: Denmark to the north; Austria and Switzerland to the south; Poland and the Czech Republic to the east; and Luxembourg, France, Belgium, and the Netherlands to the west (Worldometer, 2023).

Renowned for quality manufacturing, research, and innovation, Germany is home to global automotive leaders such as Mercedes-Benz, Volkswagen, and BMW. Globalisation has significantly benefited the German economy, with its automotive brands achieving worldwide prominence. For decades, Germany's industrial success was bolstered by access to affordable energy and labour from emerging markets (Wooldridge, 2023). The country also leads in the global footwear industry through iconic brands like Adidas and Puma, whilst retail giants Lidl and Aldi have revolutionised the grocery sector with low-price models that appeal to cost-conscious consumers globally (Funke, 2023).

Figure 6.1 The Brandenburg Gate, Berlin

Source: Getty Images

Demographic Challenges and the Future of Work

Germany's economy faces a significant threat from an ageing population, a significant demographic shift that poses a critical challenge for future talent requirements. Rural areas are becoming increasingly older and depopulated, whilst younger generations concentrate in urban centres.

In 2022, the population grew by 1.1 million (1.3%) to reach 84.3 million, a record high driven largely by net migration (Federal Statistical Office, 2023). However, the workforce is anticipated to shrink by 5 million by the end of the decade. In 2022 alone, approximately 300,000 workers reached retirement age, with the Baby Boomer generation (born 1946–1964) now exiting the labour market without a sufficient replacement cohort (Kantchev, 2021; Reiter, 2022). Below (Figure 6.2) is a 2022 graphic presentation of the ageing population. The baby boomers' period (1946 to 1964) had the most significant birth-rate increase and paradoxically, now contribute most of the ageing population. In the 1950s, in the then Federal Republic of Germany, the aged were 1 in 10. This has significantly changed as the aged population (2021) makes up 22% of the population. However, the working-age group (15 to 64) has not changed much, from 67% (1950) to 64% (2021). But what could pose a threat is the large proportion falling within the retirement bracket (1960 to 1970). Thus, demographic analysis is critical for the country's future planning and development, particularly amid dynamic changes in the workforce, workplaces, and automation, resulting in an overall future of work that is unimaginable and its implications. Germany needs to urgently address how to cope with an increasingly older population and a smaller younger population, as the proportion of the younger generation decreased from 23% to 14% between 1950 and 2021 (Federal Statistical Office, 2023).

Figure 6.2 Germany's Age Structure by Gender (1935–2070)

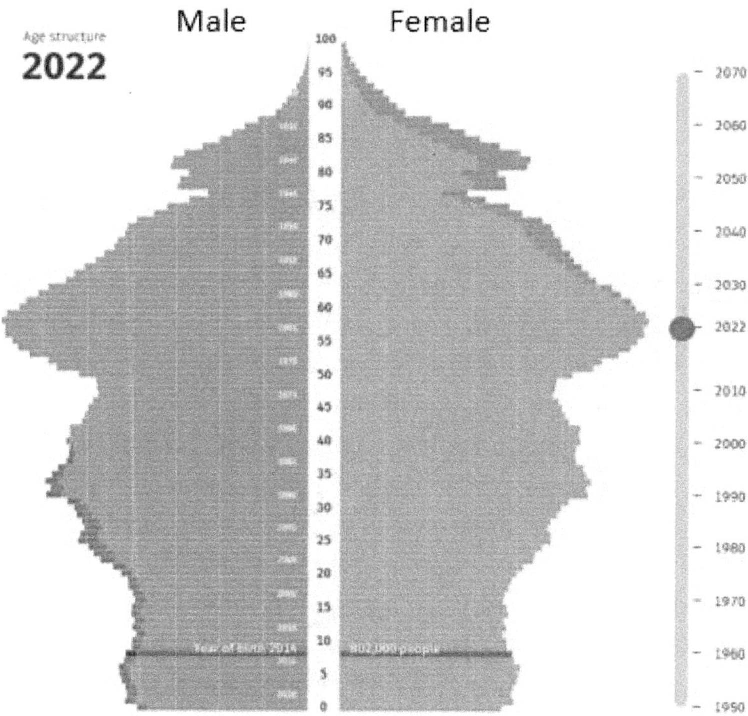

Source: Data from Federal Statistical Office (2023).

The proportion of the population aged 65 and over has risen from one in ten in the 1950s to 22% in 2021. Conversely, the younger generation (under 20) has decreased from 23% to 14% over the same period (Federal Statistical Office, 2023). This imbalance threatens the stability of social institutions and economic growth, potentially leading to labour shortages, higher taxes, and inflationary pressures (Statista, 2023).

In 2021, Germany's population was eighty-three million four hundred thousand, five hundred and fifty-four (83,408,554), a 1% increase from 2020 (Figure 6.3). Above twenty-three (23.070 million) were within the age group 40-59 (28% of the population). Age group 60 and above (24.430 million) represented (29%) (Statista, 2023), making up a total percentage population of those nearing retirement and retiring, 57% of the country's population (Figure 6.3). This could create huge challenges for future work environments, with the potential for higher prices, shortages, and increased taxes (Statista, 2023;

Kantchev, 2021), and could negatively impact the availability of workers, talent, and training for the future of work.

Figure 6.3 Germany's Population by Age Group (2021)

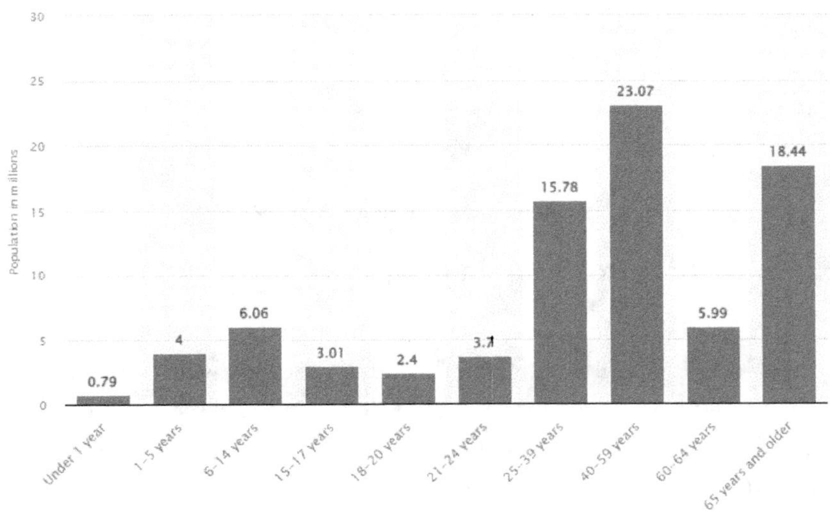

Source: Data from Statista (2023).

As shown in Figure 6.4, the population shows an increase up to 2022, but with a decline in population from 2025. The minor increases were driven by migration. The lack of significant population growth could have major implications for the future of work workforce, unless this is urgently addressed, but also could have a ripple effect on automation development and implementation. The preference for smaller families (one or two children) continues to exacerbate this trend (Kramer & Wahl, 2015; Mai, 2008; Schon & Stahle, 2019). Additionally, as women became independent, there seems to be less willingness for long-term commitment from both parties, detrimental to population growth (Schon and Stahle, 2019; Kramer and Wahl, 2015; Mai, 2008).

Figure 6.4 Germany's Population Trend (1950–2100)

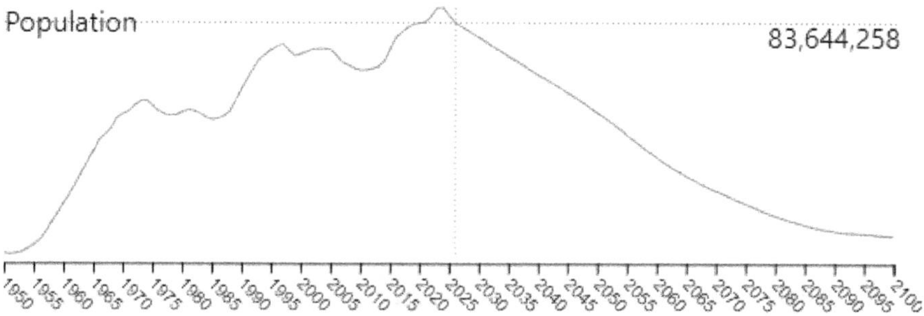

Source: Populationpyramid.net (2026)

Gender Distribution and Pay Gap

Interestingly, 2022 population of 84.3 million, had a proportion of male and female of 41.5 and 42.8 million, a 49% and 51% respectively (Federal Statistical Office, Destatis.de, 2023). The female population is higher than the male population except for the years 2013 and 2014. There is a gender wage disparity where males are paid more. This could create challenges moving into the future of work, as the younger generations expect more equality. Germany's population, when the future is rapidly changing, seems challenging due to the increase in its ageing population. Almost half the population is in the middle-aged band, effectively causing a huge concern as enormous pressure is put on social institutions, economies and businesses (PWC, 2018).

In 2021, Germany's population of 83.2 million (World Bank, 2021) was distributed by gender as follows: 42.2 million females and 41.1 million males. About 20% of German men are over 65, and 25% of women are in the same age group. Interestingly, there are more women in Germany than men, but recently, the number of men has been increasing. The future of work could mean more women in jobs. Although women make up a larger share of the workforce than men, there is a gender pay gap that needs to be addressed. Women get paid less than men, disadvantaging them in their old age earnings and pensions (OECD, n.d). Belgium is the only OECD country with the most equal pay between genders at 1.7%. The difference in South Korea is very significant, 31% (Statista Research Department, 2023).

Germany Population as of December 31, 2022, by Gender, by age Group

Within the age group 65 and above, 19.8% are men and 24% are women. Most of the population is 40 years and above. The age group (15 to 39-year-olds) is fewer than those within the age group 40 years and above. (Figure 6.5).

Figure 6.5 Germany Population by Gender and Age Group (2022)

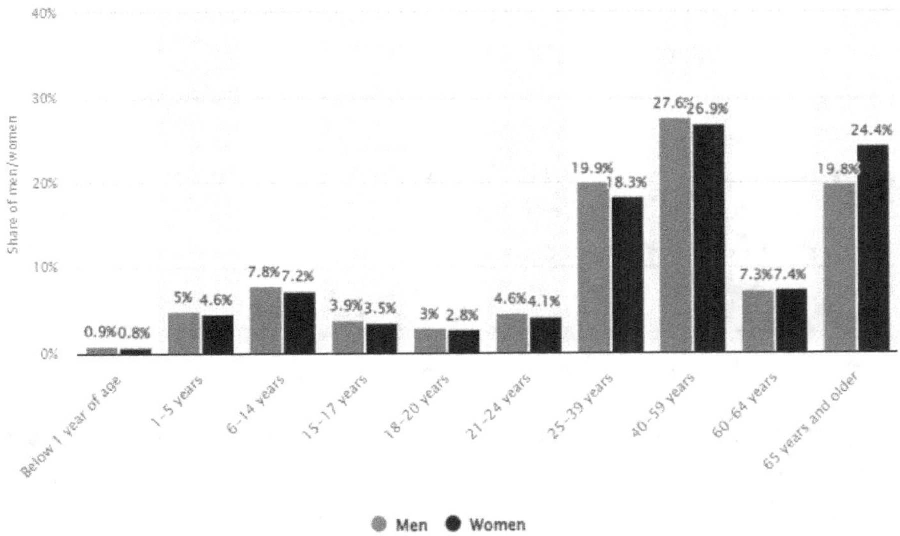

Source: Data from Statista (2023).

Socio-Ecological Market Economy and IT Infrastructure

Germany champions a 'socio-ecological market economy', prioritising social equity and environmental sustainability (Kritikos et al., 2018). This approach aligns with United Nations Sustainable Development Goals (UNSDGs) 8 (Decent Work and Economic Growth) and 13 (Climate Action). As a major trading nation, Germany plays a pivotal role in shaping sustainable global supply chains.

The transformation of work in Germany is driven by automation and digitisation. Early projections suggested that half of all jobs were susceptible to automation (Dengler & Matthes, 2018; MIT Initiative on the Digital Economy, 2017). However, the focus has shifted from substitution to augmentation, where technology complements human labour (vom Brocke et al., 2018).

The future of work and deglobalisation benefit Germany's mid-sized niche companies, whose success stems from focusing on serving the local market as the country becomes less dependent on China (Wooldridge, 2023). At risk of automation is one in eight German jobs, but this is a very small figure. SMEs employ 60% of workers and specialise in unique products, making it easier to reskill and upskill their workforce for FOW requirements driven by automation and digitalisation. According to Davies (2022), reporting on figures from a 2018 online survey, 250 German company managers reported that IT infrastructure is associated with challenges in workplace and mobility concepts, particularly in data security and data recovery programs. Similarly, a 2019 survey on German industrial companies' adoption of Industry 4.0 found that 66% of respondents were concerned about investment costs and data security and protection (Statista Research Department, 2022). Although the FOW is set for massive use of AI, automation, chatbots, and more, cybersecurity and consumer and worker safety should take precedence.

Fast forward to 2030, and the future of work is dynamic and fast-paced. COVID-19 accelerated the pace of change as digitalisation and automation shape the workplace, changing the nature of work and the workforce, and in some areas replacing current existing roles (McKinsey, 2020) WEF, 2030. Employees view the dynamic advancement of automation and digitalisation with suspicion as jobs are threatened, although AI has been reported to have certain limited capabilities, necessitating further investigation into its ability to work effectively with the human element. A shift from viewing automations and digitalisation as an end-all is required, and more emphasis should be placed on the augmentation effect (Cigdem, Meidute-Kavaliauskien, and Yildiz, 2023; World Economic Forum, 2023). Meaning that humans and machines could complement each other, thus lessening the negativity associated with advances in technology, AI, and chatbots. Structural changes involving movements between occupations and sectors could be envisaged in Germany, as outlined in the FOW and the MIT Initiative on the Digital Economy (2017).

Digitalisation, Automation, and the Future of Work

Digitalisation involves adopting business models that leverage technology to create new value. A survey of German businesses found that 43% acquired new customers and 24% anticipated increased revenue through digitalisation (McKinsey, 2020). But some products could change completely while others become obsolete in Germany FOW. A move towards the digitalisation of the workplace could align with UNSDG No-9 (industry, innovation, and infrastructure). In Germany, jobs currently performed by humans can be partially or fully digitalised to enhance productivity and efficiency, but this also brings several consequences.

Whilst automation threatens some manual roles, it also drives demand in other sectors. The ageing population has increased the need for healthcare professionals, requiring robust AI and robotic support to meet service delivery demands (McKinsey, 2023). Consequently, demand for STEM (Science, Technology, Engineering, and Mathematics) skills is projected to rise significantly.

Artificial Intelligence Barometer: Germany

The demand for AI-related skills in Germany is fluctuating but growing. Since 2019, job postings in AI-exposed occupations have grown by 62%. Notably, the Information and Communication sector leads this demand, mirroring trends in the US and UK (Figure 6.6) (PwC, 2025).

Figure 6.6 Cumulative Growth Rate in Job Postings vs AI Exposure (2019–2024)

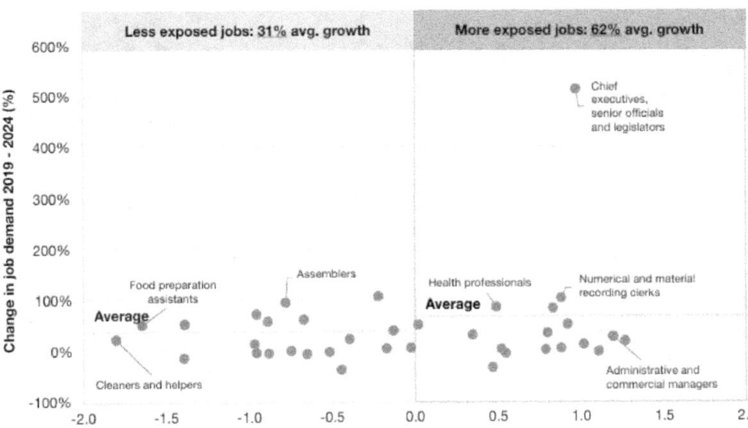

Source: Data from PwC (2025).

Germany is witnessing faster growth in AI-augmented jobs (58%) compared to AI-automated jobs (26%). The Agriculture sector leads in AI-driven job growth, with augmentation rates reaching 200%. The Education, Mining, and Recreation sectors also show high adoption rates.

Importantly, degree requirements for AI-exposed jobs are declining. Between 2019 and 2024, degree requirements for automation-exposed roles fell from 41% to 36%, and for augmentation-exposed roles from 54% to 45%. This indicates a shift towards skills-based hiring (PwC, 2025).

Industry Sectors and the Future of Work

Manufacturing

Germany's manufacturing sector, dominated by automotive, mechanical engineering, chemical, and electrical industries, employs 7.5 million people. In 2021, the sector accounted for 48.8% of exports. The government's socio-ecological policy encourages the adoption of green technologies (Kritikos et al., 2018).

Despite fears of job losses, Industry 4.0 has not resulted in mass unemployment but rather a shift in task composition. In 2022, 47% of large manufacturing companies used industrial robots. Whilst 49% of manufacturing jobs are theoretically at high risk of automation by 2030, the prevailing trend is human-machine collaboration (Davies, 2023; Owens, 2023; Shuttleworth et al., 2022).

Services and Retail

Digitalisation in the retail sector, exemplified by self-checkout systems, is reducing the need for counter staff. Similarly, routine paperwork processing in accountancy is increasingly automated (McKinsey, 2023).

In the transportation sector, autonomous vehicles threaten taxi, bus, and truck driving roles. Taxi drivers face additional pressure from the gig economy and platforms like Uber. Whilst these platforms offer flexibility beneficial for work-life balance, the potential obsolescence of driving professions causes significant workforce anxiety (Alessandrini et al., 2015; Pakusch et al., 2021).

Hybrid Work and Work Culture

Hybrid work has been associated with certain characteristics, namely boundarylessness, multitasking, the demand for constant learning, and non-work-related interruptions, which are linked to employee attitudes and well-being (Paruzel et al, 2022). The COVID-19 pandemic disrupted Germany's traditional work culture, blurring the boundaries between professional and private life (Paruzel et al., 2022). In January 2021, the government introduced the 'Mobile Work Act', granting employees the right to work from home. Different categories of working from home, namely 'Mobile' work (work that uses information technology and is carried out outside business premises, such as a coworking space) and 'Telework' (work exclusively and permanently done from an employee's home), were developed. With an internet penetration rate of 93.1%, Germany has 77.53 million people who are internet users (Kemp, 2023), making it easier to adopt any of these categories of working from home where feasible. German regulations distinguish between 'mobile work' (flexible location) and 'telework' (permanent home office setup).

Due to a strong culture of regulation, German firms implemented detailed hybrid work agreements covering data security, ergonomics, and insurance. Figure 6.7 illustrates the ergonomic standards expected for home workstations.

Figure 6.7 Ergonomic Standards for Home Workstations

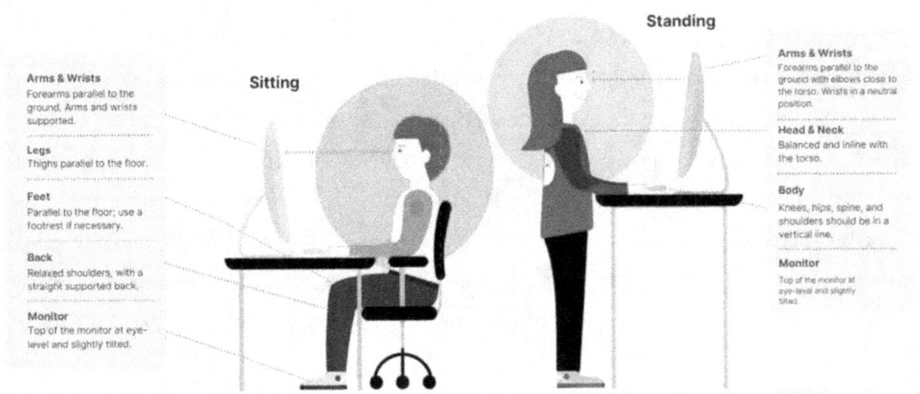

Source: Boundless HQ (2023).

Germany's Implementation of FOW Hybrid Work, Educational Qualifications, and the Industry Sector

Medium and large organisations with 100 to 1000 employees mostly implemented hybrid work policies, unlike companies with fewer than 100 employees. In other parts of the world, for instance, the UK, the size of the company does not matter. Reducing office space-related costs is one benefit for employers, while the employees' overall work-life balance could improve. Interestingly, the future of work for German younger employees is office-based, as a preference for office based confirmed by Blue (2024), contrary to those aged 45 and over, at the senior managerial level, who opt for the flexibility hybrid work offers, as the appeal for hybrid work globally remains significantly high. Germany FOW has more women than men, favouring hybrid working, as men seem to miss the office due to their social interaction, which is mainly driven by the office, whereas the opposite is true for women (Madison-bridge, 2022). Conversely, 30% of the UK's workforce works remotely, and employees aged 55 and above prefer to permanently work from home, as they have well-established, dedicated home-working spaces.

Germany's FOW and hybrid working will differ across industry sectors, with the services sector mostly adopting the hybrid model. In 2021, the automotive industry's hybrid working was at 28.4% during the pandemic, but this significantly dropped to 17.8% in 2022, suggesting not all industry sectors could continue with such a policy. But some German employers are reluctant to continue hybrid working and would prefer employees to be in the office (Madison-bridge, 2022). However, employers in the Information economy (ICT, media, and knowledge-intensive) and manufacturing are perceived as having a greater proportion adopting hybrid work as part of the future of work and an ever-changing work design and environment. Although the manufacturing sector before COVID did not have a hybrid work model, post-pandemic, a 1–2-day model is in place. Different hybrid work models apply depending on industry sector, employees, and company requirements, and vary from 1-5 days working from home.

Mostly, in Germany, large organisations plan to continue with hybrid work post-COVID, but using different models. Pre-COVID, only a proportion of firms offered hybrid work (Figure 6.8). According to a survey of 1200 German firms on hybrid work models post-pandemic, about 21% of Information economy companies had at least one employee working twice a week from home. After the pandemic, this increased significantly to 46% working from home 2 days a week (Erdsiek & Brautigan, 2022), confirming the FOW working patterns and workplaces.

Figure 6.8 Share of Firms Offering Hybrid Work Models (Pre- and Post-COVID)

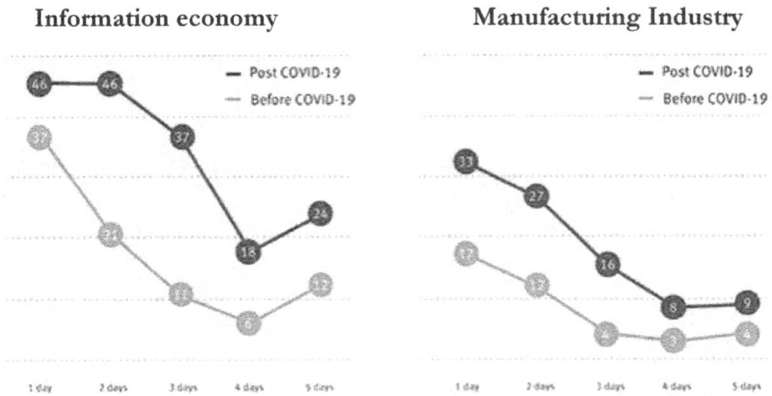

Source: Data from ZEW Business Survey (2022).

Figure 6.9 shows planned hybrid work models per organisation and organisation size. Pre-COVID, about 8% of manufacturing workers used hybrid work, but post-COVID, this is anticipated to increase to 32%, while the information economy shifts from 24% to 70% (Erdsiek & Brautigan, 2022).

Figure 6.9 Share of Firms Offering Hybrid Work Models by Firm Size

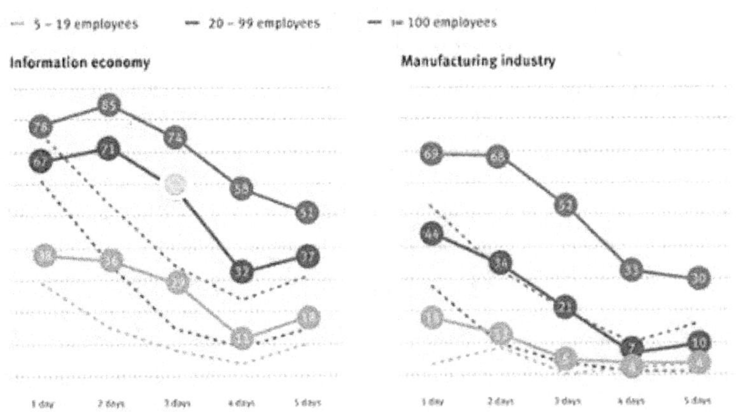

Source: Data from ZEW Business Survey (2022).

Germany's Hybrid Work Characteristics (HWC), Educational Qualifications, Employee Emotional Well-Being and Stressors, and Continuous Learning

In Germany, having an educational qualification matters for hybrid work potential. 44% of students with a university degree potentially work in a hybrid environment, compared to 22% of students with high-school qualifications, confirming that education facilitates reskilling and upskilling, as constant learning is required to keep pace with ever-changing automation and technology (Bielefeld et al, 2025; Paruzel et al., 2022).

Hybrid work characteristics (HWC) of boundarylessness, multitasking, the demand for constant learning, and non-work-related interruptions, according to Xia et al. (2019), contribute to stressors in both positive and negative ways. As Germany moves into the future of work, including hybrid work, the first three HWC are associated with job satisfaction, occupational commitment, psychological and physiological well-being, and include indicators such as emotional exhaustion and somatic health symptoms. Germany's future of work, due to rapid digitalisation, is characterised by a fast pace and multitasking. Boundarylessness and multitasking are positively associated with occupational commitment. But both have a negative effect on employee well-being, as they tend to increase somatic health issues. For boundarylessness, this does not lead to emotional exhaustion, as observed in multitasking (Paruzel et al., 2022). Constant learning is positively associated with both job satisfaction and occupational commitment. This is observed to eradicate the negative effect on emotional exhaustion and somatic health symptoms in the UK. Although the future of work is associated with reskilling and upskilling, this can be viewed as an advantage, but it also has its disadvantages when it becomes a demand for securing or keeping a job. However, German employees welcome continuous learning, and each employee is entitled to 5 days of paid education leave annually (Paruzel et al., 2022). The future of work, with its increased need for continuous learning, could bring a lot of excitement to German employees, who love learning, which in turn reduces emotional exhaustion. However, care should be taken when it involves multitasking, as it can lead to emotional exhaustion.

Working-from-Anywhere

Post-pandemic, legal rulings have upheld employer restrictions on working from abroad due to tax and social security complexities (Rehner & Braeckeler-Kogel, 2022). Unlike the 'digital nomad' policies of Spain or Estonia, Germany maintains a stricter regulatory framework, although employee demand for international flexibility remains high (Rodrigues, 2022). The future of work could determine widely suitable work locations based on employee choices, such as the US and Spain. This confirms evidence from the World Economic Forum (2023) that FOW will involve working from ANYWHERE without any designated workplaces, and the office is not considered the place to meet for work. Germany's FOW policies potentially could evaluate the future of the workplace.

Recommendations

- **Regulatory Reinforcement:** Strengthen regulations regarding health and safety for remote workers, emphasising ergonomics and the right to disconnect.
- **Demographic Strategy:** Address the ageing workforce by actively encouraging skilled migration and supporting family-friendly policies to boost birth rates.
- **Gender Pay Equity:** Establish a government task force to eliminate the gender pay gap, ensuring equitable retirement outcomes for women.
- **Trust in Automation:** Address workforce scepticism towards robotics through transparency and training that highlights the augmentation rather than substitution of human labour.
- **Human-Machine Collaboration:** Promote robotics as a complementary tool to enhance human productivity.
- **Stakeholder Collaboration:** Sustain the dialogue between government, industry, and education sectors to align skills training with future economic needs.
- **Global Flexibility:** Investigate legal frameworks to facilitate 'work from anywhere' arrangements, enhancing Germany's attractiveness to global talent.

Chapter 7

Future of Work in the G7 – France

Introduction

France, located in northwestern Europe, has a population of 68 million people (see Figure 7.2 (a)). The country shares borders with seven nations: Belgium and Luxembourg to the northeast; Switzerland, Germany, and Italy to the east; and Spain and Andorra to the south. French culture has significantly influenced the development of art and science, particularly in the fields of anthropology, philosophy, and sociology. The nation is characterised by an insistence on the supremacy of the individual; as historian Jules Michelet remarked, 'France is a person, England is an empire, and Germany is a nation' (Britannica, n.d.).

France's influence on government and civil affairs fostered democratic ideologies globally, particularly during the Enlightenment and the French Revolution, which catalysed reformist and revolutionary movements.

Figure 7.1 The Eiffel Tower

Source: Business Insider

Like many European countries, France faces an ageing population. Although the birth rate has decreased over recent years, the fertility rate remains steady at 1.79 children per woman, ranking France as one of the most fertile countries in Europe. The median age is projected to reach 43.5 years by 2050, a significant increase from 32.5 years in 1995. Although France faces an aging population but the population trend shows an increase (Figure 7.2(a); 7.2(b)).

Figure 7.2 (a): Population of France 1980 to 2030

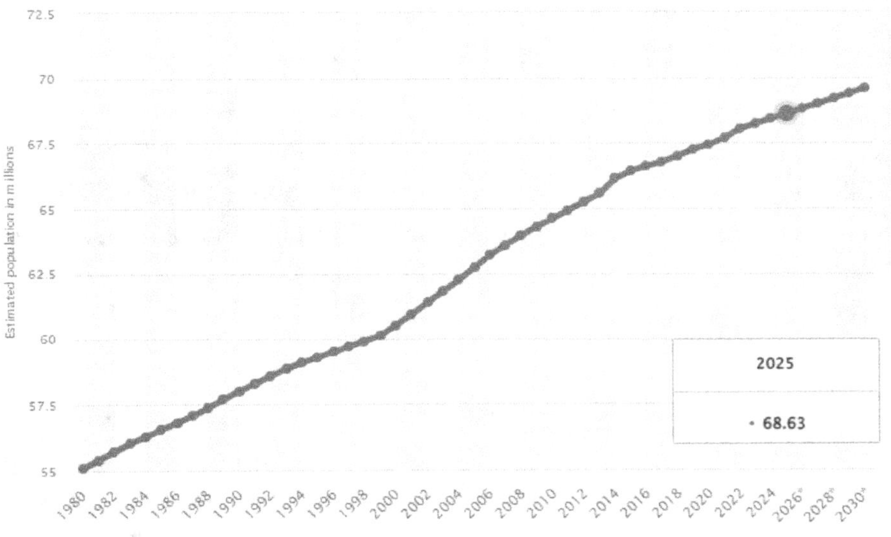

Source: Statista (2025)

Figure 7.2(b) France Population Trend 1950 - 2100

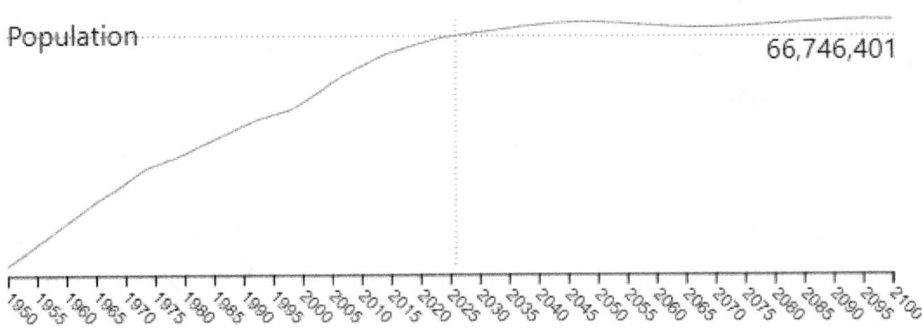

Source: Population Pyramid (2026)

Data from Population Pyramid indicates a slight decrease in population in 2050 followed by almost no significant growth for the period 2055 to 2100.

France Digital Economy and Society Index

France ranked 12th in the Digital Economy and Society Index (DESI) in 2023, with a score of 53.3. The country has progressed across several digital dimensions compared to other European Union (EU) member states (European Commission, 2024). Despite its strong position in Europe's digital economy, France lags in robotics adoption, having fewer robots than Italy and significantly fewer than German factories (McKinsey Global Institute, 2019).

The increased application of Industry 4.0 technologies threatens to widen the skills gap between skilled and highly skilled jobs. France faces critical challenges in its research and innovation ecosystem, including high youth unemployment (Suresh, 2023), educational inequality, and skills mismatches. There is widespread debate regarding the impact of digital, automation, and artificial intelligence (AI) technologies on work.

The French labour market is characterised by persistent structural unemployment. Whilst the post-pandemic period offered opportunities to boost the economy and improve employment rates, challenges remain. France invested EUR 10 billion in the 'Plan de Relance' to support employers in retaining staff, a strategy distinct from the United States (US) policy of offering direct benefits to employers (Gross & Nolsoe, 2021). These labour market issues largely predate the pandemic.

Structural Unemployment

Whilst the 'Great Resignation' and 'Big Quit' movements saw significant numbers of employees leaving the corporate world in the US and the United Kingdom (UK), a similar trend was not initially observed in France. However, recent data indicate a shift, with the percentage of workers seeking jobs increasing by 74% (Gross & Nolsoe, 2021). This record-high increase suggests a significant change in a market historically suffering from structural unemployment. As of the third quarter of 2025, the unemployment rate in France was 7.7%, compared to 5% in the UK and 3.8% in Germany (OECD, 2025).

Unemployment in France is rising faster than economic growth (Suresh, 2023), signalling a decline in productivity. The employment rate remains stable for those aged 25–55, but is low for both younger and older demographics. This structural issue is expected to persist. France has one of the lowest retirement ages in the developed world at 60.5 years. Compounding this is the demographic shift: the population aged 65 and older is increasing, whilst the 15–64 working-age population is declining (see Figure 7.3). These factors may deter corporations from investing in the learning and growth of older employees.

Figure 7.3 Employment Among Younger and Older Populations in France Post-Pandemic

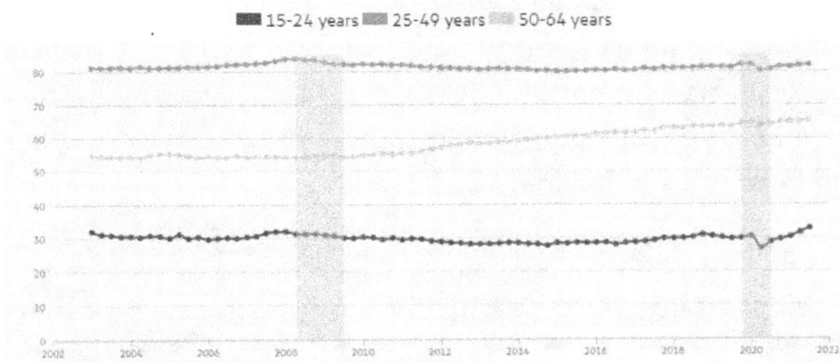

Source: Financial Times.

Automation and the Nature of Work

It is estimated that 3 million jobs in France are at risk of automation, particularly those involving high levels of manual and routine work, such as maintenance and labour. Another 15 million jobs will be partly automated, whilst 12 million will be less impacted. France is expected to witness significant skills shortages in the future, with a rising demand for profiles capable of deploying AI (Employment Advisory Council [COE], 2017). Research highlights that the skills life cycle will shorten (Organisation for Economic Co-operation and Development (OECD), 2017).

France faces the converging challenges of digitalisation, automation, heightened globalisation, and high unemployment. Technological progress and emerging technologies are transforming the nature of work, disrupting the labour market and necessitating increased flexibility and decentralisation. The future of work is shifting towards a knowledge-based economy where digital and essential skills are increasingly necessary for survival and growth.

Artificial Intelligence Barometer: France

In 2024, France witnessed a strengthening labour market with an increase in job postings requiring AI-related skills. The manufacturing sector is the leading employer, with demand for such workers rising from 13.3% in 2021 to 15.5% in 2024. Professional, Scientific, and Technical activities rank second (10% in 2024), followed by Human Health and Social Work, which rose from 4.7% in 2023 to 5.2% in 2024 (PwC, 2025).

The Information and Communication sector leads in AI adoption, with AI-related job postings increasing from 2.5% in 2018 to 3.8% in 2024. Notably, the Professional, Scientific, and Technical activities sector steadily increased AI integration in 2024, resulting in more than 2% of job postings (see Figure 7.4), overtaking Financial and Insurance activities.

Figure 7.4 France Share of AI Job Postings by Sector (2018–2024)

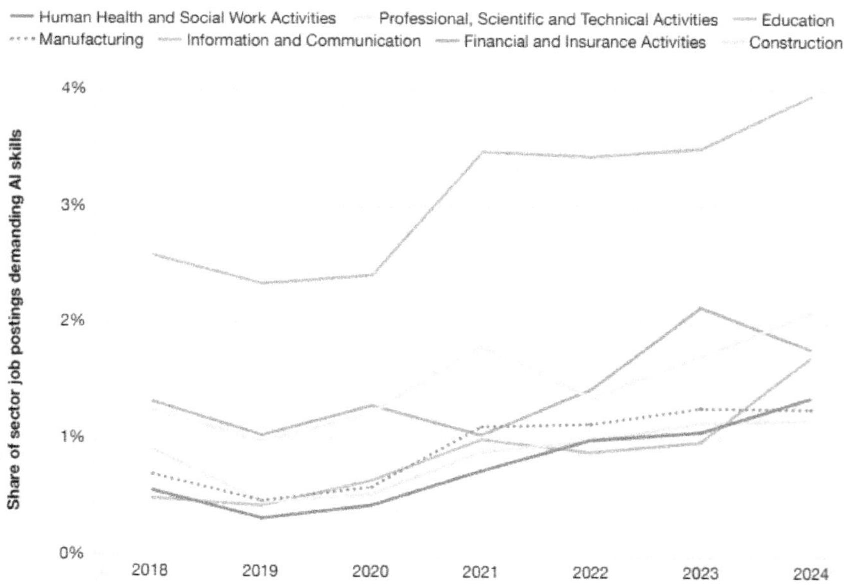

Source: Data from PwC (2025).

France has experienced significant growth in job numbers for AI-exposed roles, with a growth rate of 273% since 2019. Roles exposed to Generative AI (Gen AI) grew by 274% (see Figure 7.5). Higher AI Occupation Exposure (AIOE) is linked to faster job posting growth between 2019 and 2024. Job postings with less exposure to AI also experienced an average growth of 261%. Jobs with higher AI exposure show a positive correlation with net skill changes over the same period, indicating that AI-exposed occupations are transforming and requiring workers to upskill and reskill more frequently.

Industries exposed to AI augmentation experience higher job growth across all sectors compared to automation-exposed jobs, demonstrating a demand for employees with AI skills who can work alongside technology. Roles impacted by automation also show higher net skill changes, particularly in Health and Social Work.

Figure 7.5 France Cumulative Growth Rate of Job Postings with Gen AI Exposure (2019–2024)

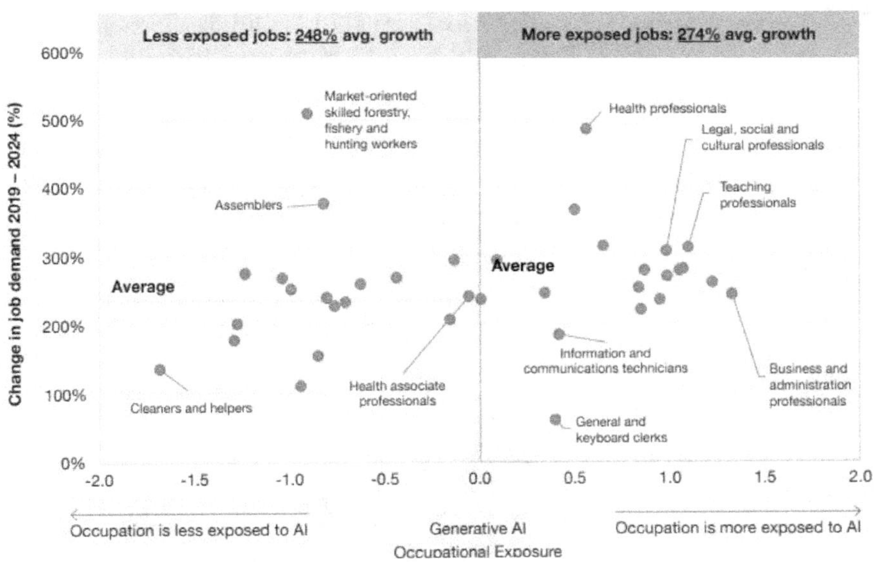

Source: Data from PwC (2025).

Degree requirements for AI-exposed jobs in France increased to 58% in 2024, compared to 54% in 2019. Conversely, degree requirements for jobs with lower AI exposure are declining. Jobs exposed to AI augmentation saw degree requirements rise to 62% in 2024 (from 58% in 2019), and automation-exposed jobs saw an increase to 52% (from 49% in 2019). Overall, formal educational requirements are increasing within both AI augmented and automated roles, demonstrating the continued importance of degrees in the French labour market (PwC, 2025).

Flexible Working and Attitude Towards Work

In France, attending the workplace is a social habit and a preferred work style pattern. Although France has struggled to embrace the work-from-home revolution, recent research indicates that remote work, or télétravail, is becoming a long-term feature (Martinetti, 2021). However, France retains a traditional hierarchical work culture. A Morgan Stanley survey indicated that post-COVID, 83% of employees in France returned to their offices (Hird, 2020).

Globally, 38% of employees now work in a hybrid mode (Microsoft, 2022), and it is predicted that 25% of all professional jobs in North America will be remote (Johanson, 2022). Whilst many nations are transitioning to hybrid work, some face technological, logistical, or cultural barriers. French employees are notably sceptical about future of work scenarios involving 'Work from Anywhere' (WFA) or a four-day work week. According to a French think tank, only 29% of French workers stated they work remotely at least once a week, compared to 51% of Germans, 50% of Italians, 42% of British, and 36% of Spaniards (Johanson, 2022). Traditional top-down management practices contribute to this reluctance.

Figure 7.6 French Telework Preference

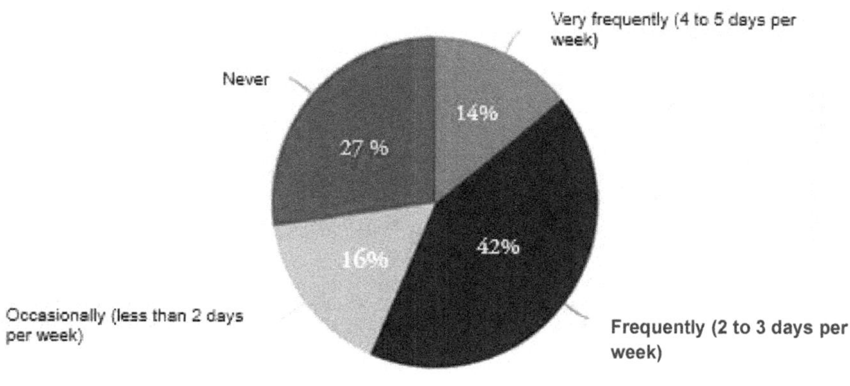

Source: Data from Statista (2025).

Despite the slow adoption, French employees express a preference for telework (see Figure 7.6). Research indicates that 42% of those surveyed wish to work remotely two to three days a week, whilst 14% prefer four to five days. However, 27% stated they would never like to work remotely, and 16% preferred fewer than two days a week (Statista, 2025).

French perspectives on work, organisational hierarchy, and work-life balance differentiate them from other G7 nations. French law restricts work interactions after office hours ('Right to Disconnect'), enforcing a clear demarcation between personal and professional life.

Recommendations

- **Youth Unemployment:** The government should collaborate with industry and education providers to develop programs aimed at training and equipping the youth with relevant AI technologies and automation skills to reduce the unemployment rate among the youth.
- **AI and automation skills shortage:** The government and industry to address the potential shortage of AI and automation skills by designing and rolling out programs to upskill and reskill the workforce.
- **Upskilling and reskilling:** To remain relevant, the workforce of the future should constantly find opportunities to upskill and reskill in line with AI and technological advancement.
- **Hybrid work pilot:** Employers could pilot hybrid work or increase access to working from home for workers for flexibility.
- **Education inequality:** The government to assess and introduce an equitable education system by widening access to all and thereby preparing and equipping future generations with appropriate education that facilitates upskilling, reskilling and retooling.
- **Right to disconnect:** The government to continue to encourage the 'right to disconnect' for work-life balance and an increase in mental wellbeing.

Chapter 8

Future of Work in the G7 – Italy

Introduction

This chapter investigates the trends, challenges, opportunities, and impacts of automation, digitalisation, artificial intelligence (AI), augmented reality (AR), and robotics on the future of work within the Italian context. The Italian Republic (Italy) is a European Union (EU) member state located in Southern and Western Europe. Geographically, it comprises a peninsula extending into the Mediterranean Sea, bordered by the Adriatic, Ionian, and Tyrrhenian Seas. The Alps define its northern mountainous boundary, and with rolling hills dominating the landscape, only 20 % of the terrain is flat (Maps Italy Geography, 2023).

Figure 8.1 The Leaning Tower of Pisa

Source: Shutterproof

As the world of work evolves, triggered significantly by the COVID-19 pandemic, working patterns in Italy are transforming. In 2020, Italian workers adopted the term 'intelligent working' to describe their rapid adaptability to pandemic-driven requirements (Ways of Working, 2020). Italian companies continue to introduce radical changes to policies, regulations, and practices, shaping a new nature of work that incorporates remote, agile, smart, and flexible patterns (Klasera et al., 2023).

Despite these shifts, Italy faces economic imbalances, including high government debt, weak productivity, and labour market vulnerabilities. Industrial activity is concentrated in the North, whilst the South relies heavily on agriculture and tourism. Although economic growth rebounded to 7% in 2021 post-pandemic, it slowed to 3.7% in 2022 (European Commission, 2023).

Industry Sectors and GDP Contribution

In 2022, the services sector was the largest contributor to Gross Domestic Product (GDP) at EUR 291,807 million, followed by manufacturing (EUR 75,677 million), construction (EUR 21,586 million), and agriculture (EUR 7,609 million). In the first quarter of 2023, Italy's GDP growth exceeded its European counterparts by 0.5%, driven by investments in construction and a recovery in tourism.

Organisations in the real estate industry rapidly adapted to new work models, incorporating hybrid and flexible arrangements that mirror emerging future of work trends. Investments in strategic workforce opportunities are shaping the transformation of this sector (JLL Global Research, 2022).

Demographics and Economic Employment Structure

Italy's population is in decline. As of January 2023, the population stood at 58.85 million, a reduction of approximately 132,400 (0.2%) from January 2022 (see Figure 8.2). By December 2023, a further decline was recorded. This downward trend is expected to persist throughout the 21st century, as death rates continue to exceed birth rates. By 2050, the population is projected to fall to 52.3 million. Additionally, Italy faces a significant challenge with an ageing population; the median age is 46.5 years (46.4 for men and 47.5 for women) (Eures, 2023; Worldometer, 2023).

Figure 8.2 Italy Population: Historic, Current, and Forecast (1960–2099)

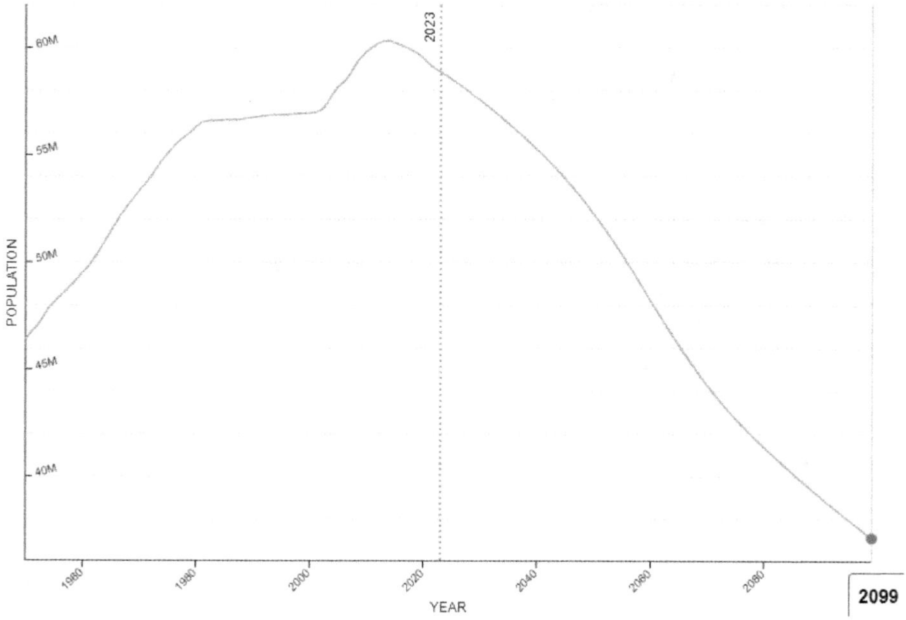

Source: Data from World Population Review (2023).

Employment, Education, and Gender

The basic educational qualification required for employment is typically an upper secondary school diploma, which secures fixed-term or full-time contracts (Eures Europa, 2022). The labour market faces structural challenges, including high youth unemployment (Suresh, 2023) and low female participation (Suresh, 2023). Globally, 11% of women's jobs are threatened by automation and AI, compared to 9% of men's jobs. In Italy, this trend could exacerbate gender disparities (Dabla-Norris & Kochhar, 2018).

Work-Life Balance and Gender Inequality

Whilst many European countries have achieved steady employment growth, Italy struggles to offer quality employment that balances work, family, and social life (Kazim, 2021). The country records lower employment levels and a wider gender pay gap than its peers. These issues are often territorial, affecting those in the South and on the islands, particularly women, young people, and foreign nationals who face instability and lower pay.

According to the Rome Business School (2022) and the Bank of Italy (2022), having more women in the workforce could increase the country's GDP by 35%. A 60% increase in the women's workforce is required to achieve the projected GDP. Both studies indicate the positive role women could play in shifting the country's GDP position in FOW. Italy's future of work and the workplace requires a balance between genders. However, a comparison of employment rates for the 20–64 age group (2019–2020) shows Italy significantly lagging behind the EU27 average (approximately 53% female employment vs 68% in the EU27) (see Figure 8.3).

Figure 8.3 Employment Rate of Population Aged 20–64 in Italy vs EU27 by Gender

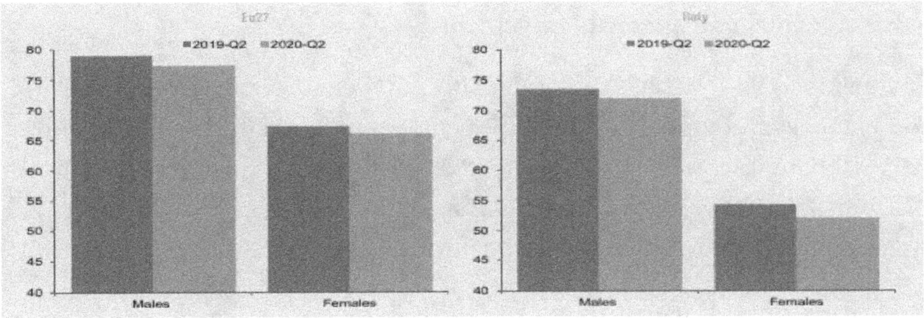

Source: Data from World Population Review (2023).

Industry Growth and Future Skills Demand

The manufacturing industry is expected to grow, particularly in transport equipment (+12.4%), pharmaceutical products (+6.5%), and refined petroleum products (+3.3%). The Excelsior Unioncamere/ANPAL Information System (2023) projects that between 2023 and 2027, Italy will require 3.8 million new workers (approximately 760,000 per year). The services sector is anticipated to absorb three-quarters of this requirement, followed by manufacturing and agriculture.

Italy's National Recovery and Resilience Plan (NRRP) aim to create new jobs primarily in construction and infrastructure (21%), tourism and trade (18%), advanced services (16%), and education and culture (13%). Conversely, sectors such as wood, paper and printing, energy supply, and mining are expected to decline.

Demand will rise for engineers, mechatronics and robotics technicians, and healthcare professionals (doctors, nurses, physiotherapists) to support NRRP-funded facilities. Information and Communication Technology (ICT) specialists will be required across all sectors to facilitate the digital transition (Eures, 2023).

Green and Digital Skills

The NRRP targets the United Nations Sustainable Development Goal (UNSDG) 13 (Climate Action). This ecological and digital transformation will create green jobs in construction (sustainable design, energy efficiency), mechanics (electric motor systems), and environmental services (environmental law, energy management).

Professionals with 'e-skills', a combination of digital literacy, mathematical/IT competence, and innovation management, will be in high demand. Estimates suggest these roles will constitute over 20% of new job requirements (approx. 875,000–959,000 jobs). Software analysts, telecommunications engineers, and systems administrators will be essential for this green transition (Eures, 2023).

Robotics and Automation

Italy boasts a robust robotics and automation industry, with 655 companies, leading Europe in this sector ahead of Germany and France. Specialising in packaging and processing (handling, picking, packing), Italian manufacturers are key partners for global sectors including food, pharmaceuticals, and personal care. Italy's growth in robotics and automation is rapidly increasing, and Italian companies serve both domestic and global markets. As Italy is faced with a declining population, there is a greater challenge of labour shortage, and this could jeopardise FOW requirements.

Despite the shrinking workforce, the growth of e-commerce and the need for productivity gains are driving the adoption of robotics. Automation offers a potential solution to labour shortages caused by demographic decline, reshaping the future of the work landscape.

Technology, Digitalisation, and Artificial Intelligence

Technological advancement will play a significant role in Italy's future of work. Whilst digitalisation and AI aim to augment human efficiency, job displacement is a risk, particularly for low- and medium-routine operational tasks. These roles, often held by women with lower educational qualifications, are highly susceptible to automation. Consequently, the technological gap threatens to widen gender disparities (International Monetary Fund [IMF], 2018; Mercer, 2022). Digital technological transformation should introduce flexibility, widen women's workplace participation, and break the glass ceiling. Policies must focus on reskilling to ensure equitable access to newly created roles.

Artificial Intelligence Barometer: Italy

Since 2018, the manufacturing sector has been the largest employer with the highest demand for workers. However, demand for AI skills has increased significantly across all sectors. The Information and Communication sector dominates AI job postings, rising from 2.4% in 2018 to 4.6% in 2024. The Professional, Scientific, and Technical activities sector follows (Figure 8.4) (PwC, 2025).

Figure 8.4 Italy Share of AI Job Postings by Sector (2018–2024)

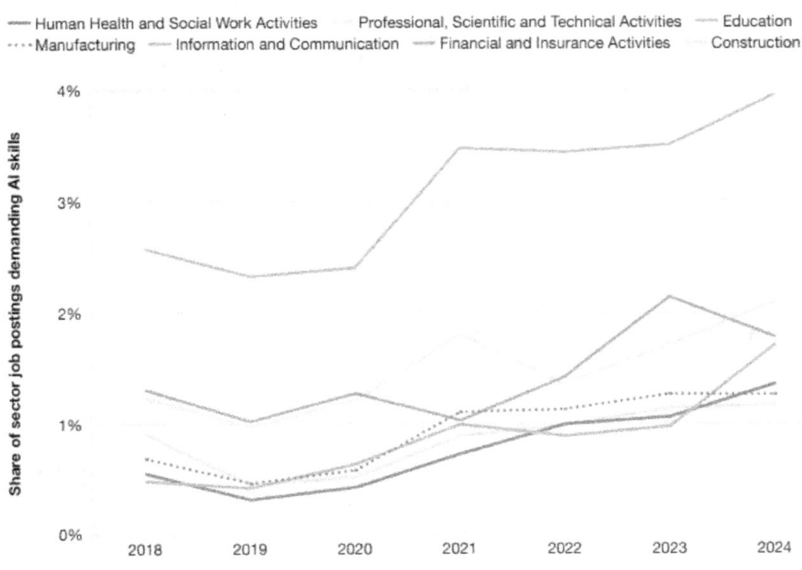

Source: Data from PwC (2025).

Job numbers for AI-exposed roles have grown by an average of 176 % since 2019. Interestingly, occupations with lower AI exposure experienced even higher growth (228 %). This suggests that whilst AI exposure transforms roles, it does not necessarily drive volume growth in isolation. However, AI-exposed jobs show a positive correlation with net skill changes, indicating that workers in these roles are upskilling more frequently to keep pace with technological change (see Figure 8.5). AI is facilitating opportunities for people without formal degrees due to lack of resources or time. The demand for formal degrees by employers is declining in jobs exposed to AI, and those roles automated by AI.

Figure 8.5 Italy Cumulative Growth Rate in Job Postings vs AI Exposure (2019–2024)

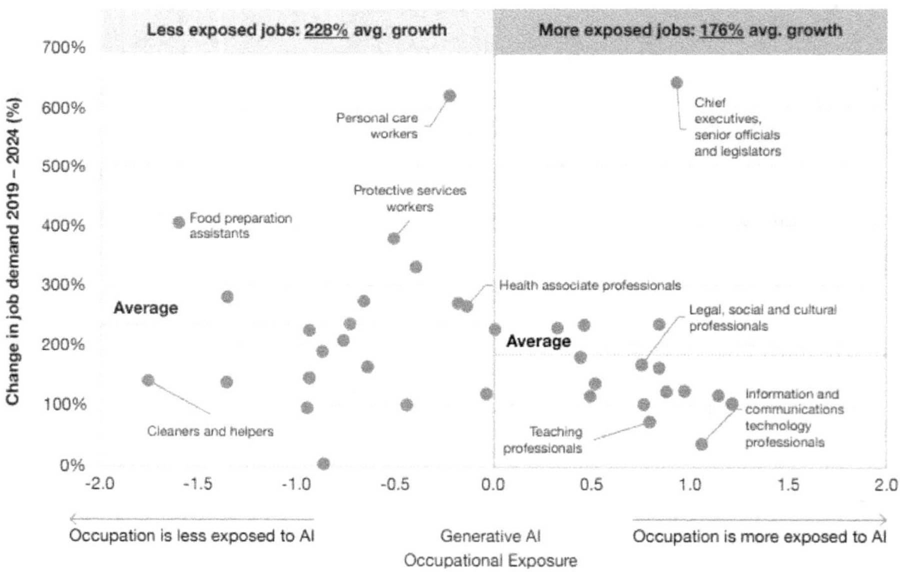

Source: Data from PwC (2025).

Future of Work Trends

Post-COVID-19, Italian companies are increasingly adopting flexible work policies. For example, Fineco Bank introduced a policy allowing 14 days of remote work per month, supporting employees with caregiving responsibilities (Kazim, 2023). Companies are creating new roles such as 'Head of Agile Work' to manage policies on time, space, and psychological support (Klasera et al., 2023).

A Socio-Technical System framework suggests that Italy requires a high degree of personalisation in work models. Hybrid work must be tailored to individual and organisational needs rather than applied as a 'one-size-fits-all' solution. Negotiated and differentiated work patterns will be essential for fulfilling future roles driven by digitisation and automation.

Recommendations

- **Skills Training:** The government should collaborate with the private sector to implement training programmes in digitalisation, AI, and machine learning.
- **Gender Inclusion:** Implement policies to increase female participation in the workforce, particularly in sectors with lower automation risk.
- **Flexible Work Policies:** Institutionalise flexible hours and hybrid working patterns to promote inclusivity and work-life balance.
- **Economic Drivers:** Investigate and support enablers for women's employment, recognising its direct link to GDP growth.
- **Youth Engagement:** Target policy measures to include a significant number of young people in the workforce.
- **Personalised Models:** Encourage personalised hybrid work models to recruit and retain talent across all sectors.
- **Employee Involvement:** Ensure employee participation in shaping future work frameworks to facilitate successful adoption.

Chapter 9

Future of Work in the G7 – Japan

Introduction

Japan, located in East Asia, is a nation where ancient traditions and cutting-edge technology coexist. With a 2025 population of over 123 million (see Figure 9.2(a)), the country is renowned for its strong work ethic, technological innovation, and high living standards. Historically, Japan has significantly contributed to the global economy, particularly in automotive manufacturing, electronics, and robotics.

However, as the country faces an ageing population, declining birth rates, and a shrinking workforce, the future of work has become a critical consideration. With advancements in automation and artificial intelligence (AI), traditional job roles are evolving, and new industries are emerging. This chapter explores the trends, challenges, and opportunities facing the Japanese workforce and economy.

Figure 9.1 Mount Fuji and Tokyo Tower

Source: Escape.com.au.

Future of Work in Japan

The future of work in Japan will be markedly different from the past. The traditional cultural outlook of 'live to work', characterised by extended hours, is shifting towards a 'work to live' philosophy. The technological revolution, accelerated by the COVID-19 pandemic, is significantly impacting work styles.

Japan faces a demographic crisis with a shrinking population and a weakening labour force. Rural areas are experiencing severe depopulation, further straining economic stability.

Figure 9.2(a) Japan population 2017 to 2027

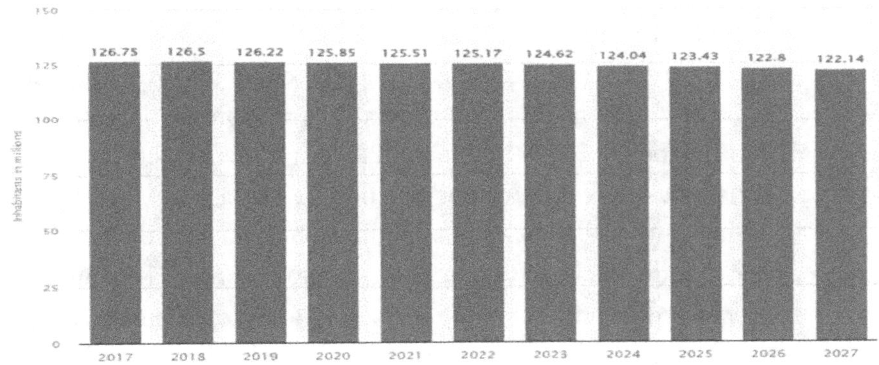

Source: Statista (2022)

Figure 9.2(b) Japan Population (1950–2100)

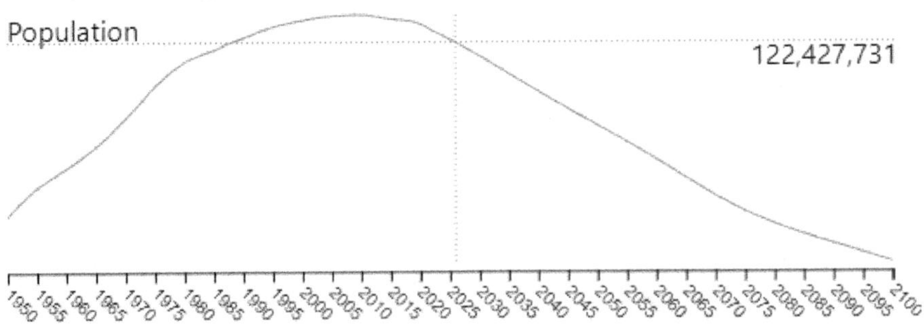

Source: Populationpyramid.net (2026)

Data from the Population Pyramid forecasts (Figure 9.2(b)) Japan's population for the period 2025 to 2100 with continuous growth, signifying that Japan's population will shrink, potentially causing future of work challenges due to a possible shortage of workers.

Technological Innovations

Technological advancements, particularly in Artificial Intelligence (AI), augmented reality (AR), and robotics, are expected to blur the boundaries of time and space by 2035 (Horii & Sakurai, 2020). This transition marks a new era for Japan, offering individual workstyle options that will transform industrial and corporate employment structures.

Japan's domestic workforce represents 59.7% of the total population. However, demographic shifts threaten to stagnate economic growth over the next decade. In 2022, the population stood at 125 million and is projected to decline to 122 million by 2027 (Statista, 2022). High life expectancy and low fertility rates have resulted in the world's highest percentage of elderly citizens (aged 65 or older) (Figure 9.3), placing a significant burden on the social welfare system, as more individuals leave the labor force. Another challenge Japan faces is that many emigrants are leaving for other countries (Statista, 2022). All these factors will create a labor supply shortage in the coming years. An ageing society and decreasing birth rate will influence Japan's economy. A research study predicts that the number of workers in the information, technology, communications, medical, welfare, and other service industries will increase, while other sectors are projected to decline. Industry growth across Japan will be skewed by the shrinking population (Horii I & Sakurai, 2020).

Figure 9.3 Japan Population Distribution (1950–2060)

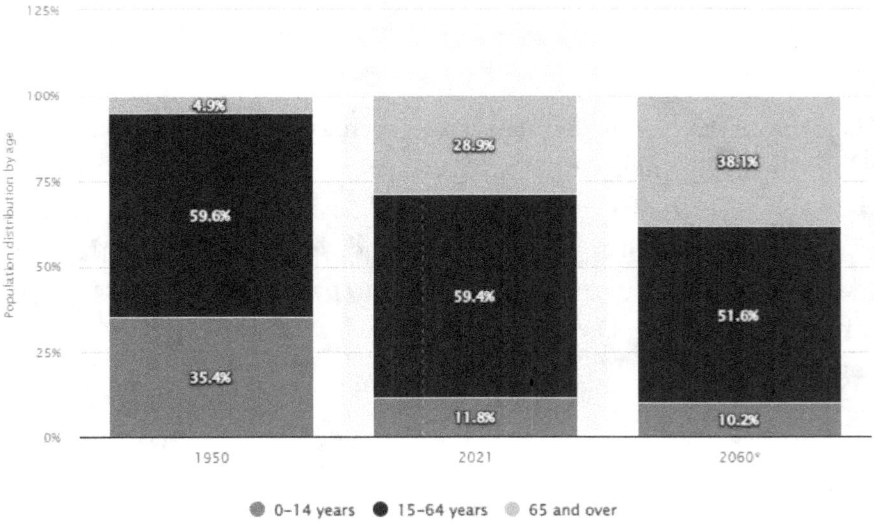

Source: Data from Statista (2022).

Technological innovation has historically driven business productivity and lifestyle. The technological revolution has enhanced communication efficiency, and manufacturing advancements have improved societal maintenance, addressed societal issues, and enhanced the overall convenience and productivity of the nation. Mobile communication currently uses 10 or 5 Gbps, and over 100 Gbps will be used by 2035, enabling widespread use of biometric sensors, virtual reality (VR), and mixed reality (MR) in business contexts.

The COVID-19 pandemic was a catalyst for Japan's shift towards digitalisation. Japanese companies are automating operations and retraining workers to deliver greater value. According to the McKinsey Global Institute, Japan plans to automate 27% of existing tasks by 2030. Whilst automation could replace 16.6 million jobs, the country will still require an additional 1.5 million workers within a decade due to labour shortages. Automation has the potential to displace 56% of work activities, reducing costs and altering work dynamics. Japan will lead the world in the potential for job displacement from automation (Horii & Sakurai, 2020).

Occupations such as payroll officers, transaction processors, and production workers have an automation potential of 69%. Conversely, roles such as CEOs and project managers are less susceptible (11%). Construction workers face a 38% risk, whilst personal caretakers face 24%.

Advancements in automatic translation technology will also benefit the economy by overcoming language barriers, potentially attracting global talent to a new ecosystem of work and living (Tirado Robles, 2022). By 2035, the workplace will be transformed, granting individuals autonomy over their working location and time. Individuals will interpret and perceive the meaning of "work" differently. The sense and definition of work will substantially change in Japan.

The Paradigm Shift

A shift in leadership mindset is essential. Organisations must focus on reskilling employees to meet new digital demands. The future workplace will be driven by an innovative culture, digitally enabled operations, and data-driven insights. There will be an increased need for 'business translators', professionals who can bridge the gap between technology and business processes. Employees with Artificial Intelligence (AI) and data science skill sets will increase. The urgent need for people to improve their digital capabilities is evident in Japan's future. In the future, companies will create an environment with enhanced flexibility. Tools and technologies will emerge that will foster a culture of team productivity and protect against cybersecurity threats.

The future of work will be characterised by flexible organisational ecosystems. By 2035, Japan is expected to see the rise of project-based organisations, where employees move between projects and organisations based on business needs. The distinction between full-time and temporary workers will blur, and multiple income streams will become common. The purpose of work will also move from only monetary compensation to goals such as social contribution.

Workers' negotiating power is expected to improve as reliance on a single employer decreases. Professional value will depend on skills mastery, necessitating continuous reskilling. However, increased virtual communication may erode the traditional sense of corporate belonging.

According to Professor Atsushi Seike, Japan will see significant structural changes in demography, technology, and market competition, which will dramatically change the future of work. Workers will be required to acquire new skills to adapt to technological innovations and structural changes (Kolade & Owoseni, 2022). To be cost-effective, companies must alter and restructure their employment practices to cope with the intensification of international competition. Technological advancements in Artificial Intelligence (AI), Augmented Reality (AR), Virtual Reality (VR), Natural Language Processing (NLP), and Robotic Process Automation (RPA), such as the humanoid robot Telexistence, will increase their application in the workplace (Malhotra, 2021). The shifting demographics will result in an acute skill imbalance and a need for more skilled workers. The pandemic experience with hybrid working has also changed the mindset of the traditional Japanese worker. The Japanese are moving away from long hours, a yes-boss attitude, and the salaried employee stereotype toward a more aware employee seeking a rewarding employment experience (Kolade & Owoseni, 2022). People focus on the company and themselves. Perspectives on career development have changed among young Japanese entrepreneurs. They are willing to quit jobs if they are not promoted quickly. Japan's lifelong employment with one company is undergoing a paradigm shift. Despite these developments and changing viewpoints, the employment structure remains rigid, and adopting flexible working practices will take time to change (Kondo, 2019).

Figure 9.4 Japan Supersmart Society 5.0

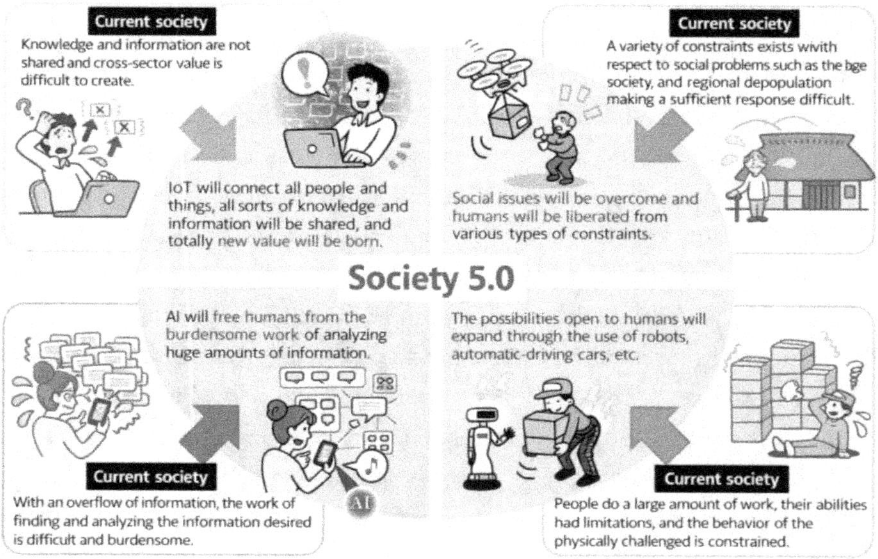

Source: CAO Japan

Digitalisation is the starting point for this transformation. Japan will witness the emergence of new business models, changes in consumption patterns, and a redesign of production, resulting in the disappearance of current work and the creation of employment and work approaches that keep pace with rapid environmental and technological advancements (See Figure 9.4). The shortage of IT professionals is projected to grow from 170,000 to approximately 780,000 by 2030. Hybrid work is gaining traction, with research suggesting it will become the 'new normal'.

Figure 9.5: Japan Percentage of AI Users and impact on work environment

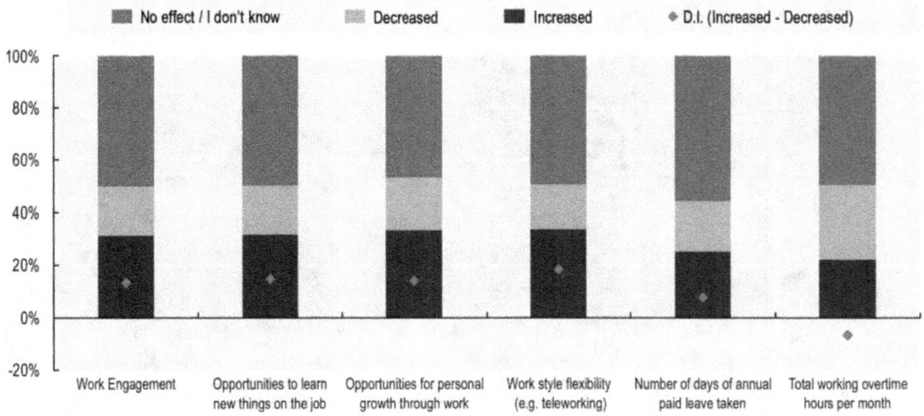

Source: OECD (2025)

Figure 9.5 shows Japanese workers' views on AI and improvement in work environment, with an overall perception of there being a decrease in total working overtime hours per month. A positive contribution to work-life balance and wellbeing.

Compared to the United States (US) and Europe, Japan remains office-oriented. However, real estate trends show a reduction in office floor space expansion (under 18%) as companies consolidate premises and adopt activity-based working. The office will likely serve as a central hub, with telework and satellite offices facilitating decentralisation. Japan will continue to see a strong correlation between office-based work and flexibility, with flexibility likely to gain traction in the coming years.

In Japan, AI users are most likely to be male (Figure 9.6), young and middle-aged, and educated. This is similar to some of the G7 countries within the OECD.

Figure 9.6: Japan Manufacturing Industry (N-682): Percentage of AI users or non-AI users by age, gender and education

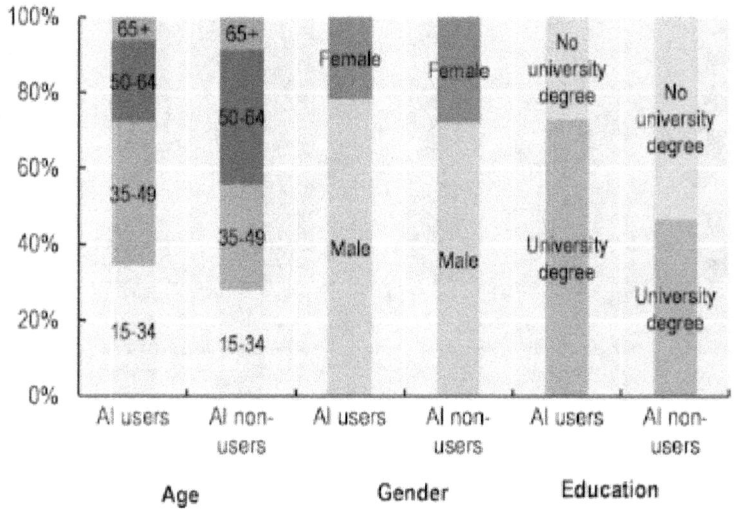

Source: OECD (2025)

Recommendations

- **Investment in AI:** The government must increase research spending on AI to address economic challenges related to population decline and ageing.
- **Sector-Specific AI Application:** Implement AI in the agriculture, forestry and fishery sectors to improve productivity and profitability.
- **Skills Development:** Government, corporations, and educational institutions must collaborate to bridge the digital skills gap between graduates and employment requirements.
- **Innovation Culture:** Encourage training that fosters creativity and innovation, as new ideas will create jobs that offset those lost to automation.
- **Inclusive Workforce:** Redesign employment practices and policies to increase the participation of senior citizens, women, and international workers in the labour market.

Chapter 10

The Interview Analysis

Introduction

Thematic analysis is a widely utilised method in qualitative data analysis, with its roots tracing back to the mid-20th century alongside the rise of qualitative approaches in psychology and the social sciences. Its popularity stems from its flexibility and applicability across diverse disciplines (Braun & Clarke, 2006; Castleberry & Nolen, 2018; Kiger & Varpio, 2020; Vaismoradi & Snelgrove, 2019). Researchers employ thematic analysis to identify, report, and document patterns, trends, or themes within a dataset. It is primarily used to analyse interview transcripts, focus group discussions, and other qualitative data sources.

Attride-Stirling (2001) posited that thematic analysis conceptualises the research process by valuing participants' recollections, thereby justifying further investigation and synthesis. The method focuses on identifying similarities and differences in repeated patterns within the dataset. Virginia Braun and Victoria Clarke formalised this approach in 2006, defining it as a method for identifying, analysing, and reporting patterns (themes) within data (Braun & Clarke, 2006). It serves as a constructionist method that examines how events, realities, meanings, experiences, and so on are the effects of a range of discourses operating within society.

Methodology

This study employed a qualitative research design using semi-structured interviews to explore perspectives on the future of work. The following subsections detail the sampling strategy, data collection procedures, analysis method, ethical considerations, and trustworthiness measures.

Sampling and Participants

Participants were identified through professional digital networks (LinkedIn) using purposeful and snowballing sampling. The search terms included 'Future of Work', 'Futurist of Work', and 'Work Futurist'. This process yielded an initial list of 50 potential candidates who were approached regarding their willingness to participate.

The final sample consisted of 22 professionals (see Table 10.1) from diverse backgrounds across the G7 nations (United Kingdom, United States, Canada, France, Italy, Germany) and other regions (e.g., British Virgin Islands). The participants included professors, researchers, coaches, project managers, C-suite executives (CXOs), chairpersons, and consultants. This diversity ensured a multicultural perspective on the future of work, representing various management levels, years of experience, and areas of expertise.

Data Collection

Data were collected via semi-structured interviews conducted on Microsoft Teams and Zoom. The total interview time was 18 hours and 30 minutes, with each session lasting approximately 35 to 75 minutes. The interview protocol comprised demographic questions followed by inquiries related to the book's central themes, including the sustainability of the future of work, long-term implications, and alignment with the United Nations Sustainable Development Goals (UNSDGs).

Interviews conducted on Microsoft Teams were transcribed automatically by Microsoft Office, and NVivo was used for thematic analysis, whilst those on Zoom were transcribed using TurboScribe.ai. All transcripts were manually checked and edited for accuracy, ensuring errors and repetitions were removed. This process resulted in approximately 250 single-spaced pages of transcripts, averaging 11 pages per interview.

Data Analysis

The study utilised Braun and Clarke's (2006) six-phase framework for thematic analysis:

1. **Familiarisation with data:** Transcripts were read repeatedly to note initial ideas.
2. **Generating initial codes:** Relevant features of the data were coded in a systematic fashion.
3. **Searching for themes:** Codes were collated into potential themes.

4. **Reviewing themes:** Themes were checked against the coded extracts and the entire dataset.
5. **Defining and naming themes:** Ongoing analysis was conducted to refine the specifics of each theme.
6. **Producing the report:** The final analysis was selected and related to the research question and literature.

The analysis was inductive, meaning codes and themes were derived directly from the data rather than fitting it into a pre-existing coding frame.

Ethical Considerations

Ethical standards were maintained throughout the study. Participants were informed of the study's purpose and nature prior to the interviews. Participation was voluntary, and informed consent was obtained. To protect confidentiality, participants are referred to by codes (e.g., P1, P2) in the data reporting, although their professional titles and locations are disclosed to provide context (see Table 10.1).

Trustworthiness

To ensure the reliability and trustworthiness of the findings, a rigorous approach to data management was adopted. Transcripts were verified against audio recordings to ensure accuracy. The coding process involved a thorough, exhaustive, and inclusive examination of each script. Themes were cross-referenced to ensure internal coherence and distinctiveness. The use of a standardised six-step method (Braun & Clarke, 2006) provided a consistent framework for analysis (see Figure 10.1).

Figure 10.1 Braun and Clarke's (2006) Six-Step Thematic Analysis Process

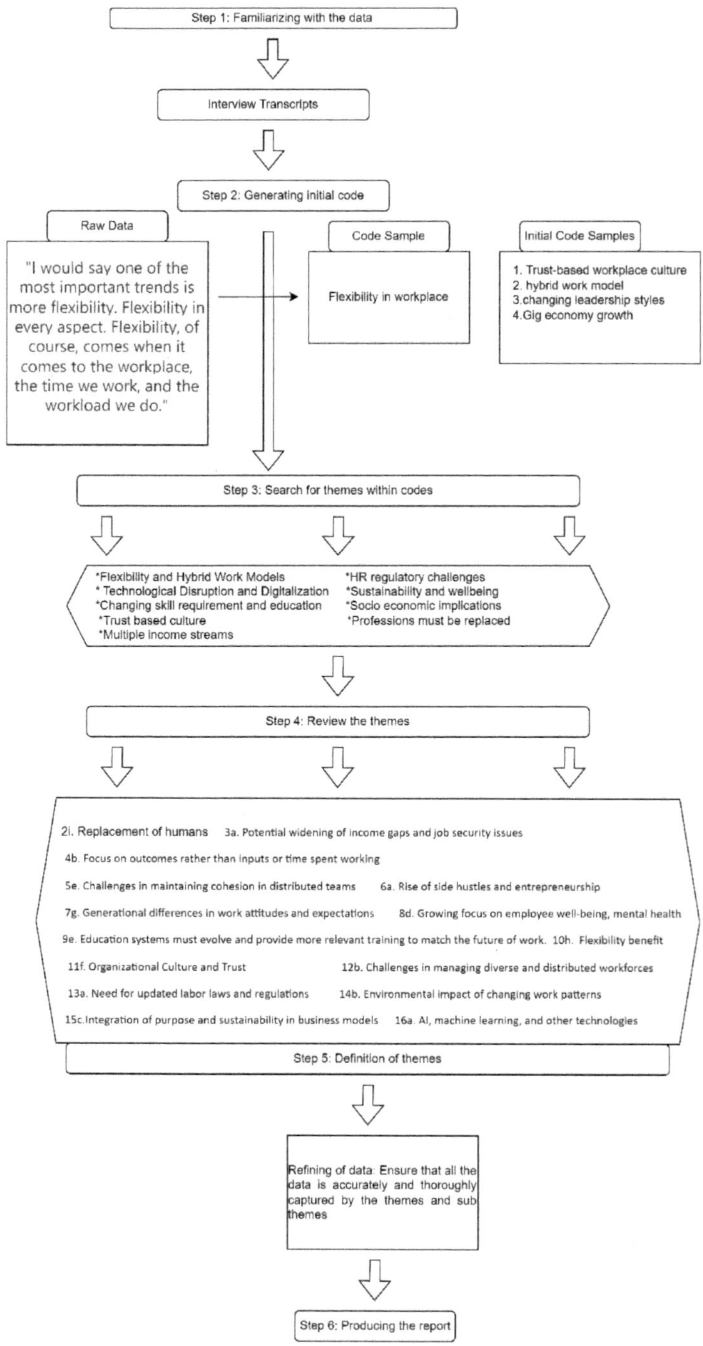

Source: Adapted from Braun and Clarke (2006).

Table 10.1: Demographics and Interview Details of Study Participants

Participant	Job Title	Gender	Country	Duration (mins)	Transcript Pages
P1	Consultant HR	Female	UK	60	15
P2	Author (Future of Work)	Male	US	35	8
P3	Researcher	Male	UK	40	10
P4	Int. Development Finance	Female	US	42	10
P5	Director of Digital Futures	Male	US/Aus	58	12
P6	Author/Educator	Male	US	50	12
P7	HR Consultant/Author	Male	UK	35	8
P8	Media & Insights Consultant	Female	UK	64	15
P9	Independent Consultant	Male	UK	48	12
P10	Consultant/Writer	Male	UK	60	15
P11	HR Manager	Male	UK	40	10
P12	CEO	Female	UK	60	5
P13	Chairman	Male	UK	55	8
P14	Project Manager	Male	Italy	59	11

Participant	Job Title	Gender	Country	Duration (mins)	Transcript Pages
P15	Chief Impact Officer	Female	Canada	75	10
P16	Professor	Male	France	45	10
P17	Professor	Male	UK	39	12
P18	HR Coach	Female	USA	60	13
P19	Speaker (Future of Work)	Female	Germany	49	15
P20	Chief HR Scientist	Male	Germany	55	12
P21	Researcher	Female	UK	38	14
P22	Professor	Female	Virgin Is.	43	13
Total				18h 30m	250

Coding and Theme Development

We examined each item of the interview scripts separately and collectively, giving equal importance to all data during coding. We did not rely on pre-existing codes; instead, we extracted relevant codes and themes from a critical analysis of the transcripts to form 16 initial themes. These were then refined to ensure coherence and uniqueness. Table 10.2 demonstrates the development of codes from raw data.

Table 10.2: Samples of Coding Raw Data

Raw Data Extract	Code
"Now, many academic jobs are moving online where academics can teach from anywhere around the world." (P14)	Remote work in academia
"So, the power is shifting from the employer to the employee regarding how work is managed..." (P12)	Shift in power dynamics
"Google and Microsoft ultimately asked their employees to work from home during the COVID-19 pandemic..." (P16)	Tech industry embracing remote work
"The contemporary view says you can be anywhere and still be very productive."	Location-independent productivity
"We are going to have a massive transformation in the future of work." (P4)	Transformation of work
"The first one that I'm noticing is transparency... employees seek more transparency..." (P15)	Demand for transparency
"The second element is flexibility. And to choose when and where they work." (P8)	Flexible work arrangements
"There's a real focus on technology... in light of the digital transformation." (P9)	Focus on digital transformation
"Remote work allows people to work anywhere... They earn a lot of money, but they work where living conditions are lower..." (P5)	Geographic arbitrage
"I think as soon as they start to rethink what they need, and then maybe downsize some of the buildings..." (P2)	Rethinking office space
"Companies have to play a role here and ensure that they have the well-being... of their employees... in mind." (P10)	Focus on well-being

Table 10.3 illustrates the alignment between interview questions, sub-themes derived from the analysis, and the final emerging themes.

Table 10.3: Thematic Structure Aligned with Interview Questions

No.	Interview Question	Sub-Themes	Theme
1	Demographics	(N/A)	(N/A)
2	Major trends of future of work?	2a. AI, machine learning, and other technologies reshape work across sectors. 2b. Automation will lead to changes in job roles and create new opportunities. 2c. Digital skills and adaptability will be crucial skills for workers. 2d. Augmentation of the human workforce with technology. 2e. Impact on job locations and hiring practices. 2f. Industry four and five technology. 2g. Automation, artificial intelligence, and Big Data. 2h. Advancement in driverless transportation. 2i. Replacement of humans. 2j. Increased use of Robotics – Avatar democratisation of work. 2k. Decoupling of work from the workplace, and work being decoupled from the G7 and moving to the E7. 2l. Democratisation accelerates the democratisation of work. 2m. ChatGPT.	Technological Disruption and Digitalisation

No.	Interview Question	Sub-Themes	Theme
3	Socio-economic implications?	3a. Potential widening of income gaps and job security issues. 3b. Changes in consumption patterns and service delivery. 3c. Impact on urban planning and infrastructure development. 3d. Implications for family structures and work-life balance. 3e. Changes in local economies and housing markets. 3f. Gender roles and equality in the workplace. 3g. Generational differences in work attitudes and expectations. 3i. Wider gap between the rich and the poor. 3j. The change from the traditional life model of I learn, I do, I retire at age 65/70, to I learn, I do, I learn, I do, and take a sabbatical rest in-between.	Socio-economic Implications
4	Are employers ready?	4a. New leadership styles and management approaches. 4b. Focus on outcomes rather than inputs or time spent working. 4c. Need for transformational and adaptive leadership. 4d. Importance of inclusivity and diversity. 4e. Challenges in maintaining cohesion in distributed teams.	Trust-based Cultures in Hybrid Environments
5	What if (employers) are not ready?	5a. New leadership styles and management approaches. 5b. Focus on outcomes rather than inputs or time spent working.	Trust-based Cultures in Hybrid Environments

No.	Interview Question	Sub-Themes	Theme
		5c. Need for transformational and adaptive leadership. 5d. Importance of inclusivity and diversity. 5e. Provide solutions to challenges in maintaining cohesion in distributed teams. 5f. Rethinking the operating model for hospitals, the role of nurses, doctors, and administration. 5g. Make work more sustainable and equitable by stretching talent, e.g., in the NHS. 5h. Freeing nurses from administrative, trivial, and repetitive tasks to use more valuable skills by stretching nursing talent. 5i. Move away from being stuck in the work operating system/model created a long time ago. 5j. Creation of a better pool of talent with skilled people.	
6	Most impacted professions?	6a. Rise of side hustles and entrepreneurship. 6b. Changing perceptions of job security. 6c. Need for individuals to manage multiple career paths. 6d. Apprenticeship opportunities reduction due to AI and ChatGPT. 6e. Accounting firms from the traditional pyramid model to the stovepipe model due to AI and ChatGPT. 6f. Education jobs – students' use of AI and ChatGPT.	Multiple Income Streams and Careers

No.	Interview Question	Sub-Themes	Theme
7	Consumption/Production effects?	7a. Potential widening of income gaps and job security issues. 7b. Changes in consumption patterns and service delivery. 7c. Impact on urban planning and infrastructure development. 7d. Implications for family structures and work-life balance. 7e. Changes in local economies and housing markets. 7f. Gender roles and equality in the workplace. 7g. Generational differences in work attitudes and expectations. 7h. Significant increase in production of basic products and services due to the reduction of the middle-income earners. 7i. Disruption of industries serving the middle class as demand profile shifts to basics.	Socio-economic Implications
8	Country-specific factors?	8a. Influence of country-specific factors on future work trends. 8b. Challenges in managing international workforces and regulations. 8c. Potential for increased global competition for talent. 8d. Growing focus on employee well-being, including mental health. 8e. Potential positive impacts on climate due to reduced commuting. 8f. Need for sustainable business practices and alignment with SDGs.	Global and Local Sustainability

No.	Interview Question	Sub-Themes	Theme
9	Implications on education?	9a. Traditional degrees are becoming less important than practical skills. 9b. Recruitment shift from degree-based to skill-based hiring. 9c. Greater emphasis on lifelong learning and adaptability. 9d. Focus on critical thinking and soft skills. 9e. Education systems must evolve and provide more relevant training to match the future of work. 9f. Making education more accessible, bite-sized, and integrated into the flow of work. 9g. Universities providing unlimited free access to ongoing university resources, education for graduates/alumni, as is being provided by Harvard Business School and the National University of Singapore.	Changing Skill Requirements and Education
10	Passing cloud or permanent?	10a. All the respondents have stressed the importance and need for flexible work arrangements. 10b. Hybrid models with remote and in-office work arrangements will be standard. 10c. The hybrid working trend model will impact work-life balance, productivity, and Organisational structures. 10d. The shift in power from employer to employee. 10e. Perceptual change in work arrangements and productivity. 10f. Working 4-day week. 10g. Hijack with hybrid.	Flexibility and Hybrid Work Models

No.	Interview Question	Sub-Themes	Theme
		10h. Flexibility as a benefit.	
11	Compensation and benefits?	11a. Shift towards contract work, gig economy, and multiple income streams 11b. Change in job security and career paths. 11c. New employee value propositions. 11.d Rise of entrepreneurship and intrapreneurship. 11e. Challenges in workforce planning and predictability for employers. 11f. Organisational Culture and Trust. 11g. Shift from traditional compensation and benefits to Personalised models. 11h. Trade off some long-term incentives for better healthcare coverage, vacation/holidays.	Evolving Employment Models
12	HR/Regulatory challenges?	12a. Need for new approaches to compensation and benefits. 12b. Challenges in managing diverse and distributed workforces. 12c. Evolving regulations around human capital reporting and transparency.	HR and Regulatory Challenges
13	Role of government?	13a. Need for updated labour laws and regulations. 13b. Support for new types of work arrangements. 13c. Collaboration between public and private sectors.	Role of Government and Policy

No.	Interview Question	Sub-Themes	Theme
		13d. The governance for work in terms of regulation.	
14	Climate change impacts?	14a. Focus on mental health and work-life balance. 14b. Environmental impact of changing work patterns. 14c. Integration of purpose and sustainability in business models.	Sustainability and Well-being
15	Health and well-being?	15a. Focus on mental health and work-life balance. 15b. Environmental impact of changing work patterns. 15c. Integration of purpose and sustainability in business models. 15d. Fundamental different mindset, fundamental different architecture for work. 15e. Designing space for learning and well-being into the flow of work.	Work-life Balance and Well-being
16	Industry/Innovation impact?	16a. AI, machine learning, and other technologies reshape work across sectors. 16b. Automation will lead to changes in job roles and create new opportunities. 16c. Digital skills and adaptability will be crucial skills for workers. 16d. Augmentation of the human workforce with technology. 16e. Impact on job locations and hiring practices.	Automation, Digital Skills, and Augmentation

Discussion of Themes

The thematic analysis involved iterative review. Initial coding yielded 99 sub-themes and 15 themes (Figure 10.2). Further refinement and the development of thematic maps consolidated these into five overarching themes that encapsulate recurring perspectives on the future of work (Figure 10.3): (Digital-First Workplace, Changing Workplace Dynamics, Evolving Work Arrangements, Adaptive Organisations, and Borderless Work.)

Figure 10.2 Thematic Analysis Map

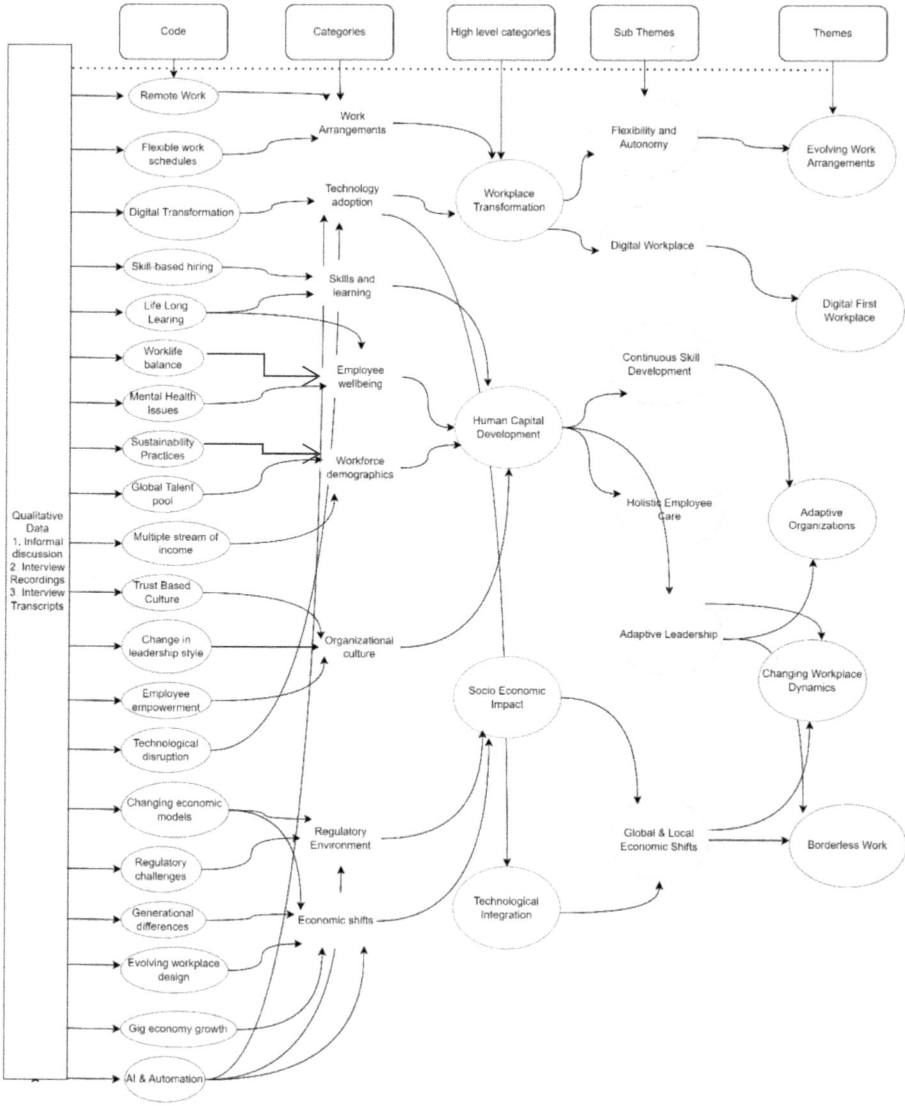

Source: Developed by authors

Figure 10.3 Finalised Subthemes and Themes

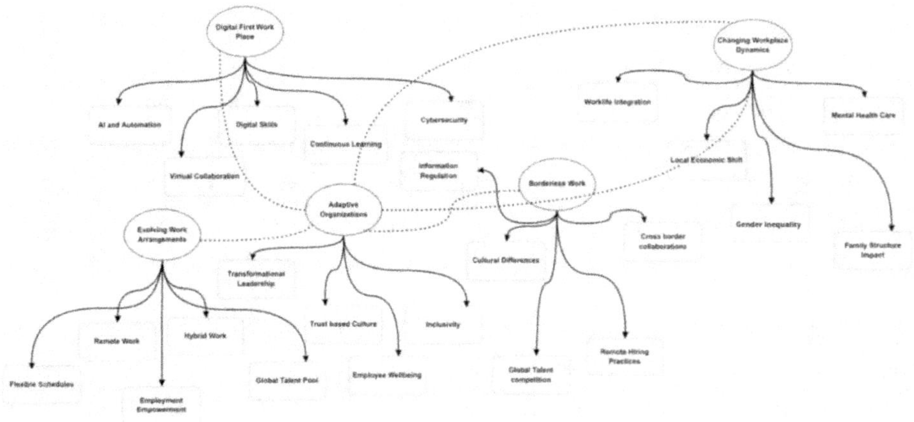

Source: Developed by authors.

Theme 1: Digital-First Workplace

This theme addresses how digitalisation, automation, and AI are fundamentally reshaping workplaces. The digital revolution is complex and multifaceted. Participants generally viewed technology not merely as a replacement for human workers but as a transformative force that augments work.

The 'Digital-First Workplace' emerged as a significant concept, highlighting the reliance on digital tools. Participant 12 noted, *'The future of work looks like it will bring us closer to using technology'*, implying that distance is no longer a barrier. This strategy drives changes such as the mechanisation of routine chores, allowing employees to focus on high-value activities. Consequently, digital skills are critical. Participant 19 emphasised: *'Technology is a big emphasis. In what ways is the organisation I work for undergoing a digital transformation? And how will they provide me with the tools to catch up and be productive rather than fall behind?'*

Conversely, Participant 2 highlighted the risks: *'A real risk of exacerbating inequality as a result of digitalisation and automation particularly, the share of wealth accruing to labour over the last 20 years is shrinking dramatically'*. This aligns with concerns regarding UNSDG 10 (Reduced Inequalities). However, Participant 2 also noted the 'democratisation of work', describing the ability to increasingly decouple work from its traditional confines…work being decoupled from the G7, and moving to the E7. This shift could support UNSDG 8 (Decent Work and Economic Growth) in emerging economies (E7 nations such as Brazil, China, India) but presents challenges for developed nations.

Virtual collaboration is vital. Participant 14 observed, *'Now, a lot of academic jobs are moving online, and academics can teach from anywhere around the world.'* This trend opens access to global talent pools. However, perspectives varied; whilst Participant 14 focused on complementarity, Participant 17 emphasised customer-oriented changes, noting tools like ChatGPT are 'totally changing customer service and content production'.

Some Participants (Participants 1, 2 and 10) acknowledge the transformative impact of technology on traditional work processes, but their views reveal a nuanced tension among professionals regarding the future of labour. Such divergent perspectives illustrate that the evolution of the digital-first workplace is not a uniform process but one that unfolds uniquely across different sectors and organisational contexts.

Participant 2 also observed two macro trends of digitalization and automation and another one called the democratization of work, and stated, *'our growing ability to increasingly decouple work from its traditional confines of space-time and structure. That means work being decoupled from the workplace, work being decoupled from the G7 and moving to the E7.'*

Interestingly, digitalization could potentially shift or move some of the work to E7 countries, namely, Brazil, China, India, Indonesia, Mexico, Russia, and Turkey which could result in the fulfilment of UNSDG 8, decent work and economic growth but the development world could negatively be affected with this shift.

As these technologies continue to advance at a rapid pace, upskilling and constant learning have become essential. To be competitive in the employment market, workers must continually adjust to new tools and procedures. Because of this emphasis on lifelong learning, Organisations are redefining their training and development strategies to provide staff members with continual learning opportunities. Parallel to these educational shifts, the rise of remote work and digital data exchange has elevated cybersecurity to a primary concern, requiring significant investment to safeguard assets and maintain the trust of both employees and customers. Ultimately, these interrelated elements of the digital-first workplace fundamentally alter how work is done, managed, and experienced across sectors, necessitating a proactive embrace of transformation for any business or individual looking to prosper in the future workplace. As remote work expands, businesses must invest significantly to safeguard resources and maintain trust.

The thematic analysis reveals that while technology and digital transformation are the key drivers of workplace transformation, their implementation and impact are neither straightforward nor uniform. Organisations must navigate complex challenges as they adapt to these changes while ensuring their workforces can effectively integrate new technologies into their practices.

Theme 2: Changing Workplace Dynamics

This theme encompasses work-life integration, socio-economic shifts, mental health, and gender equality. Participant 17 suggested workers will be 'augmented' rather than replaced: *'It doesn't mean that you're replaced by a machine or by artificial intelligence, but that you're probably replaced by someone who is using artificial intelligence... and becomes, therefore, much more efficient and capable.*" This perspective suggests a future where digital literacy and the ability to work alongside AI become crucial skills across all professions within the G7 nations. This suggests that the digital-first workplace may manifest differently across different global contexts, particularly in regions with varying levels of technological infrastructure and digital literacy.

A tension exists between technological advancement and organisational readiness. Participant 18 asserted, *'the organisations are not ready for it'*, highlighting the gap between the pace of change and adaptation.

Work-life balance is central. Participant 16 stated, *'Companies have to play a role here and ensure that they have the well-being of their employees in mind.* Interestingly, many participants (Participants 1, 7, 11, 14 & 19) have the same thoughts. Participant 22 added, *'The power is shifting from the employer to the employee regarding how work is managed, especially when work is conducted."* Many participants have said there will be a fundamental shift in workplace dynamics in favour of the employee, with the employee significantly influencing work arrangements. These perspectives highlight the significant socio-economic change in workplace dynamics, where traditional power structures are redefined. The thematic analysis clearly shows that the career path and the perception of a career have undergone a sea of change.

Career paths are becoming fluid. Participant 12 noted, *'Multiple career paths and income streams for individuals are becoming more common, fundamentally changing how people approach their careers and work-life balance.'* This perspective shows how changing workplace dynamics affect family structures and economic patterns.

Some participants perceived that the socio-economic changes would create more opportunities and foster more flexible workspaces, whereas others expressed concern that the changes would increase inequality. Participant 1 warned, *'There's going to be a greater socio-economic divide than ever before'*, a view echoed by Participant 9 regarding inequalities of voice and access to opportunity.

Multiple participants stressed that maintaining work-life boundaries in an increasingly connected workplace will become challenging and that the impact on family structure is a two-edged sword, with equal potential for more time with family and for blurred boundaries. Several participants expressed their concerns about mental health issues that would increase due to isolation and burnout. All participants unanimously agreed that changing workplace dynamics would require new perspectives and approaches to leadership and Organisational culture. The thematic analysis revealed an increased awareness and recognition that socio-economic shifts will continue to influence workplace dynamics. The participants explicitly called for improved workplace support systems, policies, and employee wellness, and emphasised the importance of work-life integration. Participants emphasised the need for new leadership approaches. Participant 2 argued, *"We are not just perpetuating what we had, but we are designing space for learning and well-being into the flow of work. I think that's going to be one of the single most important things from a socio-economic policy perspective. It's how do we design space for learning and well-being into the flow of work?"* Participant 2 added, *"And so the promises of the past are no longer relevant, and it's calling for a very fundamental different mindset, a fundamental different architecture for work and the governance for work in terms of regulation".*

This analysis reveals that while participants agreed on the significant trends shaping workplace dynamics, their perspectives on the implications and necessary responses varied by professional background, cultural context, and industry experience.

Theme 3: Evolving Work Arrangements

This theme highlights a shift in how work is conceptualised, focusing on flexibility and autonomy. Participant 9 stated, *'Work is no longer acceptable just to work and then live. But the job should enable employees to afford to go on holiday and not just once, but at least a couple of times in a year.'* Participant 20 emphasised the move from input-based to outcome-based models: *'It's not about inputs... it's about outcomes'*, highlighting a significant departure from traditional work paradigms.

However, organisational resistance remains. Participant 22 points out that flexibility is associated with shorter hours and hybrid work. Participant 3 noted *that 'employers are not ready... many of them are still living in the past'*. Participant 21 observed a divergence: *'Tech companies have to adapt quickly, but traditional companies are not in a rush'*.

The analysis and thematic maps reveal evolving work arrangements in terms of place and time of work, as well as a fundamental restructuring of the relationship between employers, employees, and work itself. The work patterns also link to digital transformation and changing workplace dynamics, suggesting that organisations must approach these changes holistically, considering both technological and human factors in their implementation strategies. Examining all the interview scripts reveals that flexibility will be the dominant trend going forward. Participant 17 predicted that the *'typical 100% work with 40-hour workweeks will be a thing of the past'*. Participant 15 reinforced the perception that people will increasingly work from varied locations; homes, coffee shops, or co-working spaces in time slots that accommodate their personal lives. The evolving work arrangements reflect the growing role of trust and flexibility. The changing work patterns call for cultural transformation with structural changes within organisations. Many participants also expressed concern as they perceive that organisations are not ready and are still living in the past. This was echoed by Participant 1, who viewed organisations as lacking a broader perspective and as focusing solely on hybrid work definitions rather than incorporating Economic, Social, and Governance (ESG) strategy, which is critical to the future of work. These perspectives reveal challenges and tension in the evolving patterns. The development suggests a significant gap between the emerging reality of work arrangements and organisational readiness to embrace them.

Theme 4: Adaptive Organisations

Future organisations will be less hierarchical, more flexible, and sustainable. All these indicators point to fluid, adaptive organisational models that prioritise sustainability and reciprocity. . Participant 19 states, *"The changing work patterns will become prominent, and all will move to contract working.*" Participant 19 further states, *"Let's say you have a company, and I'm working for you. Once this project is done, I will move to another company for another project.*" These revelations and other participants' perspectives indicate that organisations will adopt more project-based, flexible work arrangements than traditional employment models. Participant 1 echoed, *"Worker expectations have changed; it's about flexibility and work-life balance."* Confirming the need to adopt work models that fit in with employee demands and changes attributed to the future of work.

Many participants have expressed concerns about implementing and transitioning to flexible work arrangements rather than traditional employment models. Participant 2 commented on this by saying, *"You know, how do I protect the gig workers? What benefits do I provide to them? How do I handle things like employment with a company? How do I deal with independent contractors versus employees and that classification? So, I think there are many, many HR challenges."* Participant 17 states that *"the top management is convinced of the need for change, but their business revenue stream will be affected. The top management experiences challenges in convincing the middle management to accept change."* This highlights that organisations face many internal challenges in their adaptation journey with varying levels of readiness and enthusiasm. Participants' collective perspectives and opinions reflect challenges in legalities, HR concerns, and health issues. Organisations must be ready to face complex regulatory and operational hurdles. Participant 21 states that *"employees obsessed with attendance over performance will fall through the cracks. And employers who are focused on performance rather than attendance in a physical space will thrive."* These observations made by experts from various sectors demonstrate that organisations' adaptation is not just about structural changes. Still, organisations must fundamentally shift management philosophy from input to outcome-oriented metrics. A few participants shared that readiness and adaptation differ for technology companies compared to traditional companies. Participant 14 states, *"Tech companies have to adapt quickly, but traditional companies are not in a rush."* This observation highlights a two-speed adaptation process, in which different sectors move at varying paces in response to market pressures and organisational culture. This indicates that fundamental changes and the future of work arrangements are not generic but highly specific, contextual, and industry-specific. The organisation's success depends on its willingness and enthusiasm to embrace change and its ability to implement comprehensive transformations while maintaining operational effectiveness, meeting employee expectations, and maintaining competitive advantage.

The analysis reveals that adaptive organisations can be agile, balance competing demands, uphold transparency and trust, focus on outcomes, and navigate local and global constructs. Thus, they will thrive in the evolving future of work. The interview analysis also reveals that organisational adaptation is intertwined with cultural transformation, further complicating the adaptive process.

Theme 5: Borderless Work

Borderless work represents a shift in geographical distribution. Participant 16 predicted the growth of the 'hub economy', where work is independent of location. Participant 4 envisioned a future with minimum physical infrastructure but more innovation by saying, *'I want to envision a future of work in our society with minimum physical infrastructure but more of innovation.'* Most participants complemented each other on the point that digitalisation is, and will continue to play a key role in enabling borderless work (Participants 14, 18, 21 & 22). The interview analysis reveals the customer-oriented nature of borderless work. Borderless work will redefine the significance of various stakeholder groups of businesses. There is a perception among all participants that when organisations discuss the future of work, the conversation centres on digitalisation as the key enabler. However, Participant 10 emphasised that diversity and inclusion must go beyond hiring: *'You can't hire people who are different if you're not going to allow them to be different'*. Participant 19 noted that borderless work introduces 'complex regulatory, legal, and practical challenges'.

Some participants share emerging insights on the fluid nature of project-based work across Organisations. Participant 20 states, *"For example, let's say you have a company, and I'm working for you, and once this project is done, I will just move to another company for another project."* This indicates a fundamental shift from traditional employment models to more flexible, project-based arrangements that transcend Organisational boundaries. Project-based fluidity allows for new engagement models, such as Unilever's 'Flex' system mentioned by Participant 1, where employees can *'work for six months somewhere else and come back'.*

Additionally, Participant 1 argued, *'You know, and people's expectations about skills being rewarded and working differently, working gig is coming in. People can be a full time employee and then their life circumstance can change and then they can actually go.'*

The analysis asserts that the world will witness a substantial transformation in how work is organised and performed, and its implementation will unfold significant practical, legal, and cultural challenges and consequences that will vary significantly across different global contexts. The interview analysis reveals that though Organisations and professionals are welcoming and embracing borderless work, the ramifications and challenges of implementation require careful consideration and navigation of regulatory, technological, cultural, and Organisational factors.

Summary

The thematic analysis of 22 in-depth interviews reveals that the future of work is driven by complex, interconnected transformations. Five major themes, Digital-First Workplace, Changing Workplace Dynamics, Evolving Work Arrangements, Adaptive Organisations, and Borderless Work, illustrate that whilst technology (AI, automation) is a key driver, human factors remain central. Organisations must balance digital transformation with well-being, adapt to flexible arrangements, and navigate borderless operations. Successful adaptation requires a holistic approach considering technological, cultural, and organisational factors while remaining responsive to regional variations and industry-specific needs. This comprehensive analysis provides valuable insights for various stakeholders of society, such as Organisations, policymakers, leaders, and individuals, who are preparing for the evolving future of work landscape.

Chapter 11

Future of Work Perspective Through the Lens of STEEPV

Introduction

This chapter examines the key factors and trends shaping the future of work through the lens of the STEEPV framework. STEEPV is a strategic foresight tool valuable for investigating complex topics by analysing Social, Technological, Economic, Environmental, Political, and Value-based domains (Jaizuluddin et al, 2025; Loveridge & Saritas, 2009; Loveridge, 2002). This multidimensional approach clarifies the intersecting influences driving the future of work. The framework highlights perspectives from industry experts and human resource (HR) leaders, supported by secondary research.

Empirical data collected from 22 participants, primarily from G7 nations, reveal that whilst the future of work manifests differently across regions, commonalities exist. Jobs are being redefined, necessitating workforce upskilling and reskilling.

Statistics reflect a significant shift towards remote work. Approximately 40 % of the global workforce is expected to work remotely in the future, compared to 28% in 2023 and 20% in 2020. According to the World Economic Forum (2025), 91% of employees prefer fully or mostly remote work, citing improved work-life balance and mental health benefits. Furthermore, 83% of employees report successfully transitioning to remote work. However, a disconnect remains, as 83% of Chief Executive Officers (CEOs) anticipate a full return to the office by 2030.

STEEPV Framework Application on Future of Work Trends

Social

Socially, the future of work is undergoing drastic changes from both employer and employee perspectives (World Economic Forum, 2025). Workers recognise significant workplace shifts but often lack understanding of the underlying drivers. Whilst fear of job loss exists, there is a readiness to embrace and upskill in artificial intelligence (AI) and Generative AI technologies. Professions most impacted by automation include legal services, investigative journalism, marketing, accounting, and teaching. However, new AI-associated roles are emerging, absorbing the upskilled workforce (Liu et al, 2025). This aligns with United Nations Sustainable Development Goal (UNSDG) 4, which advocates for inclusive, equitable quality education and lifelong learning.

The COVID-19 pandemic accelerated the shift to flexible working (Hobbs, 2021; Ng et al, 2021), extending privileges previously reserved for senior management to the broader workforce. Whilst this shift initially caused isolation issues, the flexibility to integrate work and life has become a widely accepted norm across G7 nations.

Key social factors include the demand for a holistic work-life pattern, prioritising health and well-being (UNSDG 3). Employees in France, the United Kingdom (UK), and the United States (US) increasingly view the traditional five-day, 9-to-5 office model as outdated, as work can now be done from anywhere due to technological advancements.

The gig economy is expanding, particularly among Generation Z (Gen Z) in the UK and the US (Kurian and Bindu, 2024; Tangalakis-Lippert, 2025; Vadavi and Sharmiladevi, 2024). Flexibility is becoming central to work and life, challenging rigid work styles. The focus is shifting from viewing a job as the sole priority to valuing restfulness and a slower pace. The concept of 'micro-retirement' (Mariappanadar, 2025; Robinson, 2025; Simon, 2025) is emerging, where individuals take career breaks to travel or explore before returning to the workforce, rather than waiting for traditional retirement at 65 to enjoy life (The Week, 2025). Thus, some employees would be swapping work calendars for plane tickets and traveling back-to-back through cities and countries. FOW proposed trends could help meet United Nations SDG 3, as workers would not have to feel stressed or work under pressure while balancing work and life, since work could be more accommodating. 80% of the remote workforce have reported better mental health. Japan has seen a 28-30% adoption rate of remote work. Globally, North America and Europe are leading the way in future work trends (World Economic Forum, 2025).

Table 11.1: G7 Work Patterns and Cultural Dimensions

Country	Workweek Hours	Paid Leave	Work-Life Balance	Hofstede: Power Distance	Feedback Style	Legal Protections
France	35	5+ weeks	Strong	High	Direct, argumentative	Extensive: 35-hour law, strong enforcement
Germany	40	4+ weeks	Strong	Moderate	Efficient, precise	Robust: paid leave, overtime, parental leave laws
UK	37	4+ weeks	Moderate	Low	Positive, precise	Minimum leave, flexible work regulations, parental leave laws
US	40+	None (standard)	Weak	Low	Positive, fast-paced	Limited: Family & Medical Leave Act (FMLA) only, state-level variation.
Japan	40+	2+ weeks	Weak	High	Indirect, formal	Moderate: granted but unused, overtime laws
Canada	38	2-4 weeks	Moderate	Low	Collaborative	Varies: labour code, provincial differences
Italy	40	4 weeks	Moderate	Moderate	Formal, relationship-based	Strong: EU rules, enforced paid leave

Source: Adapted from Hofstede Insights and national labour laws.

Table 11.1 illustrates workplace expectations across G7 nations, analysing cultural dimensions such as power distance alongside legal protections. A strong work-life balance correlates with robust legal frameworks (as seen in France and Germany) rather than just shorter workweeks. In contrast, the US and Japan, with weaker legal protections for rest, exhibit weaker work-life balance despite varying cultural norms.

Table 11.1 presents both similarities and differences in workplace expectations across the high-income G7 nations. The table presents the Hofstede (1980) cultural dimension power distance, and discusses its interplay with work norms, work-life balance, feedback styles, legal structures, and organisational practices. The table shows that a strong work-life balance is not determined solely by short weeks but is also associated with legal protections, cultural norms, and workplace practices. For instance, despite their differing power distance rankings, both France and Germany achieve high levels of work-life balance. The statutory and enforcement mechanisms of both these countries support a stronger work-life balance. In France, the high-power distance, legal protection, and shorter weeks imply that employees' personal time is structurally protected even within hierarchies. Conversely, Germany's moderate power and reliance on collective agreements foster predictable boundaries and stable working hours. In stark contrast, the United States and Japan both exhibit a weak work-life balance; their legal frameworks allow for 40-plus-hour weeks and offer insufficient protection for rest time, despite differences in their paid leave structures. The low power distance, coupled with weak laws, and the high-power distance with strong hierarchy, drag down the work-life balance. Meanwhile, Canada, the United Kingdom, and Italy occupy a moderate band, where low-to-medium power distance and medium protection policies result in a balanced, though not superior, quality of life. While Hofstede's dimensions can be observed in workplace interaction, it is the underlying legal and statutory frameworks that primarily dictate the feasibility of work-life balance.

Socially, as is evident, the future of work will not be as it was before, with people working in offices or designated workplaces. Instead, a balance of agreeable workplaces and workstations will emerge. However, it is anticipated that there will be a generational difference in the workplace in terms of work attitude and expectations, with the younger generation's commitment and perception of work differing from work being viewed as an end and not vice versa, as previously viewed by the older generations. Deloitte Global Gen Z and Millennial Survey (Faber, 2025a) reveals that younger generations prioritise flexibility, mental well-being, mentorship, and meaning and financial security

They challenge traditional norms, for these generations, the future of work is based on what they think it should look like and what they want, rather than what the employer prescribes, and often engaging in 'job hopping' or 'side hustles'. Their focus on values and continuous reinvention suggests that employers must adapt future work environments to attract this talent (see Figure 11.1).

Figure 11.1 Generation Z and Millennials Prioritise Soft Skills Development

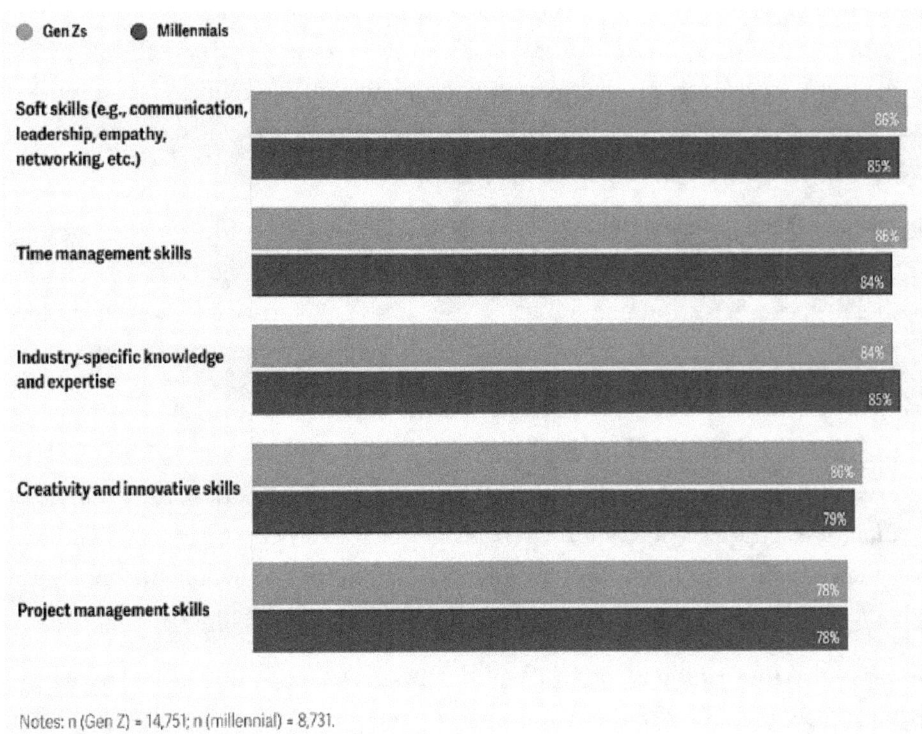

Source: Data from Faber (2025a).

Technological (UNSDG 9)

The future workplace is leaner, smarter, and technology-enabled. Hybrid models–typically two to three days remote–are becoming the norm. With 44% of companies globally integrating AI into remote settings, productivity has reportedly increased by 35-40%. This leads to 62% of employees feeling more productive from home (Guliyev, 2025; World Economic Forum, 2025).

Remote work is reshaping economies by reducing emissions and transportation costs. It is anticipated that new jobs will be created, although challenges such as the digital divide and cybersecurity risks persist (Guliyev et al, 2023; World Economic Forum, 2026).

The USA is already calling employees back to the office, despite some of the G7 nations using hybrid as the norm for FOW. 'Digital-first' strategies are central, with technology fostering interconnectedness. However, advancements in AI and machine learning will displace some roles, necessitating upskilling (Filippucci et al, 2025; World Economic Forum, 2025; Guliyev et al, 2023). Technology will increasingly be used in product design and manufacturing, reducing the required workforce. Nevertheless, the 'human-in-the-loop' remains essential to intervene in case of systemic failure (Liu, 2025).

Robotics and chatbots are expected to assume roles in healthcare, such as robotic nurses or companions for the elderly. However, advanced AI is viewed with caution due to potential ethical lapses and 'hallucinations' in systems like ChatGPT. The potential for intrusion and lack of human oversight raises significant ethical concerns.

Some driverless motor vehicles have already posed challenges due to the lack of intervention in the event of uncontrollable situations. The restrictions on charging points for electric cars pose another challenge for consumers and increase demand for infrastructure capital, particularly for organisations pursuing technological advancements, know-how, and research and development (Muratovski, 2026).

What is also interesting is the FOW and its impact on housing, as more people are choosing smaller houses or shared accommodation to reduce their environmental impact. Therefore, generally, the housing infrastructure could significantly reduce the impact on climate change and the environment. Although 22.8% of the US population works remotely, there is a rise in the return-to-work mandate among tech and finance companies. As more companies embrace digital and automation advancements, the main challenge remains the lack of evaluation of the implications of this significant shift in AI use (World Economic Forum, 2025).

Another interesting finding is that the 2024 global workforce hopes and fears survey, which covered 56,600 workers across 50 countries, revealed that 61%

of workers have used AI tools like ChatGPT and DALL-E at least once in the last year (Figure 11.2), but far fewer are using them regularly (PWC, 2024). It is to be noted that percentages shown may not total 100 due to rounding and omission of 'don't know' responses.

Figure 11.2 Frequency of Generative AI Tool Usage by Employees (Past 12 Months)

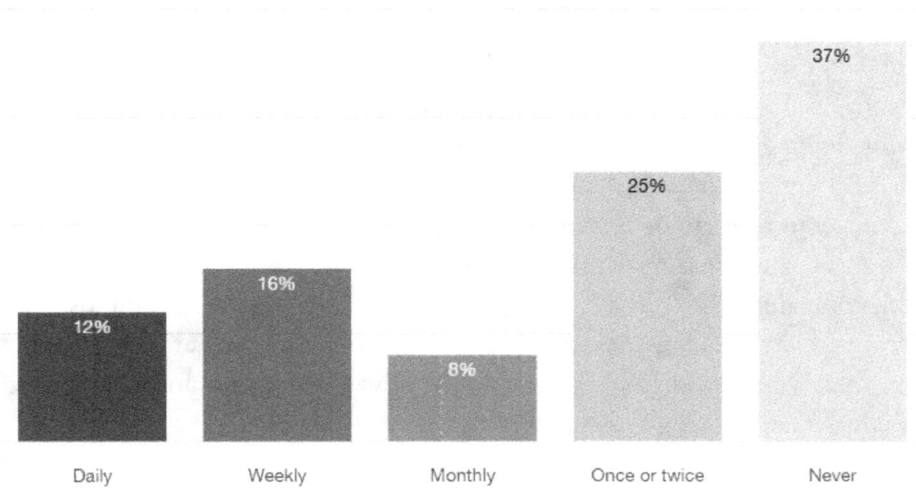

Source: Data from PwC (2024).

Economic

The income gap will widen between the rich and the poor, potentially leading to a massive rise in job insecurity. This could contradict the United Nations SDG 1 (no poverty). Advancements in technology, such as AI and ChatGPT, could exacerbate the situation. The result is changes in consumption patterns as more consumers' spending power is reduced, along with a decline in the standard of living (van Eerd & Guo, 2020). Industry disruptions will occur as companies shift to producing more basic products and services due to declining consumer affordability. Industries' supply chains could change as production line implementation shifts to basics (Liu, 2025). This could result in reduced profits and margins for companies.

Generally, economic factors create two categories of consumers: the rich and the poor, with the majority falling into the latter, as the middle class erodes due to shifts in job availability, incomes, and generational changes (van Eerd &

Guo, 2020). Thus, the straight line of the normal course of life, 'I learn, I do, I retire at age 65/70' changes to 'I learn, I do, I learn, I do, and take a sabbatical rest in-between'. There are potential challenges for companies and industries due to reduced demand for luxury products and a shift in consumer preferences towards basic products and services.

Environmental (UNSDG 13)

The UN Sustainable Development Goal 13 (climate change) emphasises the urgent need to combat climate change and its impacts by 2030. Based on the study findings, companies and consumers' awareness of ecology and the physical environment has increased. Consumers and the wider community require companies to demonstrate plans to formulate and implement measures to mitigate adverse climate impacts. There will be increased consumer demand for companies to provide tangible proof of their contributions and accountability for climate issues. This could involve reconfiguring the supply and value chain, becoming more directly involved with communities, engaging in ethics dialogues, and including activities that demonstrate a shift towards climate change awareness, such as using climate-friendly and recyclable raw materials, reusing, and minimising waste (Osei et al, 2026).

However, companies could potentially weigh the benefits of initiating and implementing climate change programs versus the costs. Large Organisations such as Shell and BP seem to have shelved some of the climate-mitigation programs they initially launched as costs mounted (Moore & Wilson, 2025).

As work becomes more democratized and evenly distributed across the world due to advances in digitalization, this could create an opportunity to reduce poverty by widening universal access to education through digital and automation platforms (UNSDG 4). Subsequently, the quality of work might improve.

However, the increase in automation and digital usage through the adoption of AI, the use of servers that fuel such technology, could potentially create more negative climate change impact as carbon emissions are emitted into the atmosphere, as well as the requirement for substantial water usage (Kenny et al, 2025; Walker and Goldsmith, 2026; Xia et al, 2025). Therefore, although there could be climate change initiatives, the widespread use of AI could have adverse effects due to increased server demand and the infrastructure required for implementation. This could be a global challenge as AI gets rolled out across the continent.

FOW requires leaner infrastructure, avoiding the presence of large houses or having accommodation in multiple locations, as it does not support sustainability. More people will own one house, with a greater focus on smaller homes rather than large ones. There could be no need for physical offices, as current offices could potentially be turned into accommodation. The sharing economy will grow with cars and driveways being shared for sustainability. Reduced consumption could help achieve UNSDG 12, which aims to ensure sustainable consumption and production patterns (Eckhardt et al, 2019; Puschmann & Alt, 2016; Schor, 2014).

Political and Legal

Governments play a critical role in the future of work, particularly regarding AI regulation. Given the high potential for AI adoption in G7 nations, governments must ensure transparency, equity, and safety. Policies are required to safeguard humans from potential AI-induced harm. While the buzz is on automation, AI, and ChatGPT's widespread use, the foremost factor should be the implementation of appropriate policies and regulations to safeguard humans (Gordon, 2021; Cath, 2018).

Whilst companies are responsible for upskilling and reskilling, governments must ensure fairness in these programmes. Collaboration between government, industry and the workforce is essential. Governments that also learn from ethical principles should draw definitive boundaries of responsibilities, particularly when it comes to the application of AI and automation (Gordon, 2021).

Demographic Shift and FOW

The demographic shift in the UK, with an ageing population, means a greater proportion of the workforce will be dealing with this segment of consumers, within health, holidays, the banking industry, or other sectors. The demand for physiotherapists, acute medical staff, and geriatric specialists may increase due to an ageing population. But there could be a shortage of such professionals. Robotics might be used to have a robotic physiotherapist or nurse, or, in some instances, a robotic pet. Lonely older adults could have a robotic friend to chat to. But the discussion on the right to live, right to die, poses challenges as Christians value life as a gift from God. Generally, this becomes a heated topic, with solutions mainly focusing on reducing the burden on health services from the ageing population. This dilemma of an ageing population cuts across most of the G7 nations, the USA, Japan, and Germany (World Economic Forum, 2025).

There is a need for a clear regulatory framework to address demographic (Liu et al, 2025) issues and protect the ageing and aged. There may be a need for personnel to monitor quality and standards, which may require establishing a professional body. The UK government could review the regulatory framework governing the recruitment of non-resident UK health sector workers to ensure transparency and that correct procedures are followed. FOW sees significant growth in the health services sector.

Non-Geographic Location of FOW

From a regulatory perspective, the demand for flexible work without geographical boundaries challenges the formation and implementation of work and employment contracts. Where workers and companies had a geographic location, FOW trends blur these boundaries, as work can potentially be done from anywhere. Governments' revenue collection could become cumbersome with non-geographically specified locations for employers and employees.

The anticipated growth of the gig economy requires a clear, specific legal and regulatory framework, which may face hurdles. Technology and automation enable FOW to be more standardised, internationally accessible, and more connected technologically not physically. Companies are expected to implement safety regulations as employees work in geographically dispersed locations. Currently, there are only national regulations, as global regulations for companies do not appear to exist. Overall, change is anticipated in FOW. The change could be imposed by geopolitical factors, AI, climate, or other forces that could require FOW changes (Gordon, 2021; Cath, 2018).

Values

Automation and technology are enabling new value systems. The hybrid working model fosters self-accountability and has been linked to increased productivity. This model allows employees and employers to participate in a value chain enhanced by technology.

New values will emerge as employees seek a higher quality of life, prioritising well-being and experiences over the traditional 'business of work' (Christian, 2023). The 'human-in-the-loop' concept remains critical for ensuring a trusted and ethical technological environment. Consumers will develop new cultures and values that shun employers and those brands that exhibit climate greenwashing tendencies, as their knowledge and awareness of climate issues and mitigating factors increase. This could result in brands that are not climate change advocates losing customers and business as the future of work develops. Emerging employee values will scrutinise employers' ethics and sustainable practices by evaluating fairness in AI deployment, leadership attitudes, data protection, transparency, and integrity (Gordon, 2021; Cath, 2018). Thus, the emerging values in the future of work will be dominant factors for choice of work, type of work, how to work, and the need for consistently upskilling, reskilling and retooling to remain relevant in order to meet the technological advanced skills needs and deployment for employers.

Reflections

This chapter has utilised the STEEPV framework to analyse the key factors shaping the future of work. Automation and digitalisation are set to transform jobs and organisations. Governments must coordinate these changes, drive climate initiatives and ensure ethical standards in AI adoption. Ultimately, the future of work will be defined by the intersection of technological advancement, political environment, social shifts, and evolving human values.

Thus, the human-in-the-loop is critical for a smooth, trusted, and ethical FOW advanced technological atmosphere. Employees will seek a higher quality of life, showing an interest in leisure and pursuing new experiences over the 'business of work'. New values will emerge as FOW trends shape how people work, manage their health and well-being, and acquire the latest skills demanded by jobs, driven by advancements in predictive analytics and AI.

Chapter 12

Concluding Remarks

The future of work is not fixed; it will continue to evolve with advancements in technology and automation, whilst being shaped by natural events and pandemics. Work models have shifted significantly from the traditional 9-to-5 schedule. The COVID-19 pandemic awakened the world to new possibilities for working models. However, there is currently no consensus amongst employers, employees, and institutions regarding what these models should look like.

In the future of work, artificial intelligence (AI), technology, and automation will play significant roles, though the degree of their involvement remains a subject of debate. Much remains to be explored, discovered, developed, and implemented. There is no longer a 'one-size-fits-all' solution.

The definition of 'work' has undergone numerous modifications, dating back to Adam Smith in 1776 and evolving to contemporary definitions (Gregory & Valenti, 2021; Thompson & Warhurst, 1998; Wrzesniewski & Dutton, 2021). These descriptions have guided the transition to the 'future of work', with the anticipation that further definitions will emerge as work continues to evolve.

This study examined the future of work across G7 nations, the United Kingdom (UK), the United States (US), Germany, Italy, Canada, Japan, and France, using horizon scanning to identify potential threats, risks, emerging issues, and opportunities. A qualitative research approach was adopted, involving interviews with influential experts, company leaders, human resource (HR) professionals, thought leaders, and labour market strategists to gain deep insights into the subject.

Drivers of Change

Significant disruptions, including technological advancements (automation, AI), the 2008 financial crisis, the proliferation of social media, and heightening geopolitical risk, the COVID-19 pandemic have all contributed to the current debate on the future of work. The pandemic, in particular, forced a global rethink of work strategies, leading to the rapid implementation of remote and hybrid working models.

Effective hybrid work incorporates flexible workspaces that facilitate collaboration and teamwork. Research suggests that hybrid work is viable and that employees seek flexibility to balance work, life, and well-being. This has resulted in employee empowerment globally, challenging traditional forms of work.

Employees increasingly demand control over their work-life balance, often viewing work as a means to an end. This trend is particularly prevalent amongst the younger workforce (Generation Z and Millennials). Conversely, some employers are mandating a return to the office, leading to friction. Employees are increasingly willing to quit if flexibility is denied, or they engage in 'quiet resistance', continuing to work in ways that suit their lives. This shift supports United Nations Sustainable Development Goal (UNSDG) 3 (Good Health and Well-being).

Skills Gaps and Socio-Economic Divides

A noticeable skills gap is developing as automation and AI are rolled out. Emotional intelligence and social skills are becoming critical. During the 'Great Resignation' in the US, major reasons for quitting included lack of well-being support and limited flexibility. Future work models must strike a balance between employer and employee expectations to avoid failure.

The future of work is anticipated to widen the divide between the rich and the poor, potentially contradicting UNSDG 1 (No Poverty) and UNSDG 10 (Reduced Inequalities). Automation and digitisation may erode the middle class, leaving many able to afford only basic products and services, which could impact organisational revenues.

Country-Specific Insights

United States of America

The US is faced with a dynamic shift from a focus on formal educational degrees to work-based models to meet with relevant and specific workplace skills demands due to advancement in automation. The country is preparing its workforce for an AI-driven environment. However, the gap between the rich and the poor will widen and the demand for basic products and services will increase. There has been an increase in the number of jobs with less AI exposure since 2019 to 2024 in industries such as health associate professionals, cleaners, food preparation assistance, similar to the trend in the UK.

Notably, degree requirements are decreasing in AI exposure jobs since 2019, with a significant decline in 2024 (from 62% to 53%). Less AI exposure jobs recorded a slight downward change in formal degree requirements.

Canada

Canada's population is growing, but AI and automation may disrupt the workforce. A significant number of senior citizens participate in the job market, often in the gig economy, to supplement retirement income. Most Canadians favour remote work and flexibility. Canada has redefined job categories, moving away from 'white-collar' and 'blue-collar' distinctions towards categories such as knowledge workers, personal services, and technical trades.

In Canada, jobs with high AI exposure experienced a decline in degree requirements. Similarly, jobs with lower AI exposure and the degree requirements declined, but with a lesser percent. This indicates a falling requirement for formal degrees but with a more focus on skills – upskilling and reskilling.

United Kingdom

The population of the UK is smaller than pre-Covid although it is projected to increase in the near future. The UK is involved in AI, robotics and other 'smart automations', bringing benefits to the economy. The UK is undergoing three waves of automation, with Waves 1 and 2 already taken place. The last phase, wave 3, is focused to take place in the mid-2030s. Women's jobs will face greater automation within the next decade, with finance jobs at higher risk in the short term. Transportation industry automation will take place over a longer time. However, the country will face an aging population challenge. Hence, some companies within the health sector have started to use robotics. Success in the implementation of AI and automation requires government and corporate involvement.

The UK will experience a generational workforce in 2030, with the involvement of older and international workers to ease this pressure. Jobs will become insecure and younger workers will use Social Media to influence each other towards a common work behaviour approach of 'doing very little'. The gig economy will expand and drones will be introduced to compensate and improve efficiencies in future of work within some industry sectors such as construction. The use of drones could increase output, GDP, profits and pay as employees upskill to meet drone technology requirements. But some sectors' wage bill could decrease due to the need for a reduced work force due to drone technology.

In UK, formal degree requirements in AI exposure jobs is falling. Generally, jobs with less AI exposure, degree requirements are also declining but at a slower pace than in AI exposure jobs.

Germany

Germany, a hub for innovation, faces an ageing population but leads in climate initiatives and social equity. The country has benefited from deglobalisation by promoting local industries. Although Germany is one of the other advanced technological nations, the introduction of digitalisation and automation might result in some products becoming obsolete. But the advantages of acquiring new customers and increased revenues could surpass the disadvantages as new products are introduced.

The government is actively involved in the implementation of policies to safeguard its people in light of the evolving work patterns. Digitalisation and automation are prominent in retail and services, though this may lead to job losses in transportation. Germany has implemented restrictions on 'working from anywhere', requiring employees to be within the country.

In Germany, employer degree requirements in AI exposure jobs are declining and the trend is also evident in AI less exposure jobs, but at a slower pace.

France

France faces an ageing population but maintains a high fertility rate. The country ranks well in the Digital Economy and Society Index. France did not experience a 'Great Resignation' but saw an increase in job seekers. The unemployment rate remains higher than in the UK and Germany. France leads in AI adoption in the Information and Communication sector. In 2024, the country experienced an increase in job postings requiring AI -related skills. Industries exposed to AI augmentation reported higher job growth compared to those jobs facing automation. Degree requirements for AI-exposed roles have increased, whereas those with less AI-exposure have decreased for the period 2019 to 2024.

Italy

Italian workers coined the term 'intelligent workers' due to their rapid adaptability to new work strategies brought about by COVID-19. The Italian work's associate hybrid work with different working patterns, agile, flexible, remote and smart. Some companies introduced radical adjustments to work patterns through policies, regulations and practices redefining future of work. At the same time, the nation faces economic imbalances and a declining population. The National Recovery and Resilience Plan places emphasis on creating green jobs. Some Italian companies operate both domestically and internationally, and the nation leads Europe in robotics and automation. However, there are jobs at higher risk of automation, particularly those involving low and medium-routine operation tasks, and clerical. The nation's goal of technological transformation is to bring flexibility, widen women's workplace participation and break the glass ceiling. AI-exposed jobs are growing across all industries in Italy, requiring workers to upskill.

In Italy, AI is democratising opportunities for people who lack the time or resources to obtain formal degrees. Employers' demand for formal degrees is declining, specifically in jobs exposed to AI, and those more highly automated by AI.

Italy's hybrid work model will be personalised in line with individual needs to meet work-life balance. There will not be a 'one-size fits all' approach.

Japan

Japan's future of work will introduce more flexibility, shifting away from traditional models. The ageing population poses challenges. Japan aims to automate 27% of existing tasks by 2030 to address labour shortages. Routine tasks are at high risk of displacement.

Japan is the only country among the G7 nations that is adopting AI at a slower pace. The government is concerned of AI risks, and hence, focusing more on introducing an AI deployment and adoption regulatory framework to safeguard both the workforce, and the economy.

Three key aspects are reshaping the future of work in Japan: aggressive automation, labour shortage, and a rapidly aging population. Japan is also gradually experiencing a shift in perceptions of work. The workforce is leaning towards a flexible work culture rather than overwork. Due to demographic pressures, companies are employing older people, women, and foreign talent. Automation is increasing Japan's productivity, which will help it to maintain GDP growth.

Thematic Analysis and STEEPV Framework

Twenty-two future of work experts involving strategists, HR consultants, chief experience officers (CXOs), researchers, directors and educators from the G7 nations, were interviewed. Purposeful and snowballing sampling techniques were applied. The interviews were analysed using the six-step thematic analysis process (Braun and Clark, 2006) and a number of subthemes and key themes emerged (Figure 10.3). The key themes were extensively analysed to arrive at participants' deeper perspectives and insights into the future of work.

Technological disruption and digitalisation were the major themes that developed from four major trends of FOW subthemes. The subthemes pointed out trends such as automation, AI, machine learning, augmentation, replacement of humans and others. The human-in-the-loop is a term that was mostly used as an advancement in technology to shape FOW to mitigate any shortcomings. Socio-economic implications were discussed and with issues such as widening of income gaps, job insecurity, generational differences at work, impact on urban planning and infrastructure development, and changes in consumption patterns, gender roles, among others.

Most employers were not ready for FOW and have not yet built trust-based cultures needed for remote and hybrid work environments, even though it is crucial for FOW. New leadership styles and management approaches, like transformational and adaptive leadership with active promotion of diversity and inclusion, are important for FOW strategies. It was widely discussed that there will be a noticeable and critical need for upskilling and reskilling as recruitment for job requirements shifts from requiring formal educational qualifications, such as degrees, to skill-based training. More accessible education provided in bite-sized chunks was indicated as part of the democratisation of education and job requirements, as well as universities offering lifelong learning facilities to their graduates. Critical thinking and soft skills become increasingly important due to evolving jobs and a new skill set in FOW.

Mostly, participants agreed that FOW was not a passing cloud but was an emerging reality. FOW is not stationary but dynamic in line with advancements in technology, economic shifts, disruptions, and geopolitical risks.

The success of FOW requires a strong HR regulatory framework and synergy between the government, industry, education sector, and other stakeholders of society. The FOW requires coordinated efforts to tackle climate issues and at the centre of the FOW, new strategies and developments, is employee wellbeing and work-life balance. The one-size-fits-all work model is no longer applicable going into FOW. There are also major implications on industry innovation, sustainability, automation, augmentation, AI and digitalisation. Careful implementation of automation, AI, augmentation would result in fewer job losses, but the need for more upskilling and reskilling to match is required with FOW newly developed roles.

Social, Technological, Economic, Environmental, Political and Values (STEEPV) framework was applied to present key factors and trends on FOW in conjunction with applicable UNSDGs (1, 2, 3, 4, 5, 8, 9, 12, and 13).

Both employer and employee will witness drastic changes and with some employees lacking the knowledge to grasp employers' FOW strategies, particularly where AI and AI-augmentation is involved. Some jobs are most at risk of displacement due to automation (legal, marketing, accounting, teaching, recruitment consultant, journalism, business development and others). Automation is the biggest disruptor in jobs and is taking over roles that involve repetition or are routine. Of the G7 nations (Figure 11.2), Japan is leading in automation of jobs as this seems to surpass AI-exposed jobs at an alarming pace. More jobs could be replaced, resulting in increased unemployment, which could be in contradiction of some of the UNSDGs (1, 2 and 8 – no poverty, no hunger, decent work and economic growth), as the unemployed might experience poverty and hunger due to lack of decent work and a shrinkage in economic growth. Introducing Gen AI to augment humans could lessen job displacement. Japan's scenario could be necessitated by the aging population as organisations aim to align future of work workable strategies for a nation that could face huge shortages of future of work skills and talent.

The automation trend is at a slower pace in other G7 nations, with Italy almost at the same pace as AI, but AI is slightly overtaking automation (Figure 12.1). This could result in the avoidance of rapid job disruptions and losses. It could also facilitate upskilling and reskilling of the workforce.

Figure 12.1 G7 Automatable Roles and Skills Disruptions

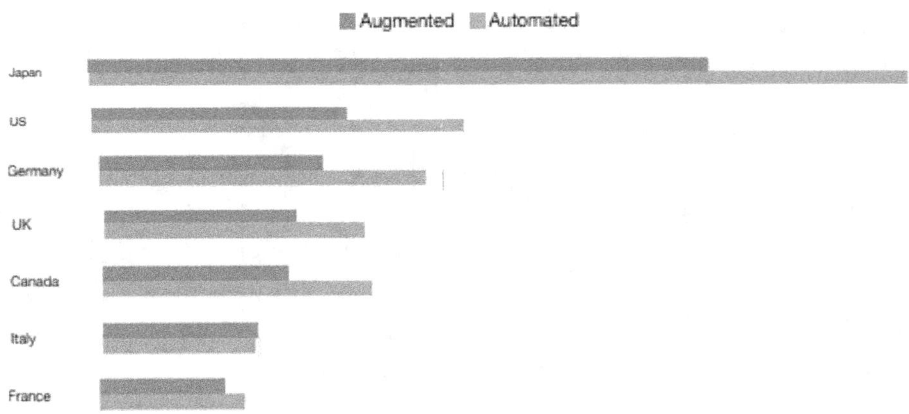

Source: Adapted from PwC (2025).

In some of the G7 nations, AI-exposed jobs are transforming and requiring workers to upskill and reskill more frequently. Therefore, AI-exposed occupations have a positive relationship with net skills changes, while those in low AI exposure require a smaller net skill change.

The gig economy is focused on expanding as it offers flexibility and a more manageable work-life balance, as the 9-5 pm traditional job models are shunned. There could be an increase in micro-retirement or sabbatical leave, where the need to experience life takes centre stage over work until retirement (UNSDG 3 – Good health and well-being).

Across the G7 nations, there are differing work week hours, paid leave, work-life balance, and feedback styles. Surprisingly, although France has a higher power distance, the working week is shorter and a more direct and argumentative feedback style could also indicate why several protests are taking place there. The three countries, Germany, Japan, and the US, have long working hours.

The widening of the income gap could increase poverty (UNSDG 1 – no poverty). It is anticipated that consumption patterns could change and lean more on responsible consumption and production (UNSDG 12). This could be linked to technology, how some products could be made obsolete and the demand for luxury products could be reduced, hence a general reliance on basic products and services. UNSDG 13 requires knowledge and action on climate issues and increasingly, FOW could result in more stakeholders embracing solutions to mitigate further adverse climate change issues.

Politically, governments are expected to play a major role in enabling policies and legislation governing FOW. Consumption patterns as well as climate change initiatives could be formulated and implemented while at the same time ensuring the reduction of poverty and the promotion of innovation in industries leading to sustainable infrastructure, thus aligning with some UNSDGs (9 - industry, innovation and infrastructure, 12 – responsible consumption and production, and 13 – climate action).

The promotion of new values of self-accountability while increasing productivity should be at the centre stage in FOW. Advanced technologies like AI and AI-augmentation are tools that could enable improved work values and work locations. This will provide a holistic approach to work-life balance while at the same time enabling the human-in-the-loop to new technological innovations to reduce job losses.

In Conclusion

Overall, the requirement for formal degree qualifications is declining in some G7 nations, signalling a shift towards skills-based hiring. Automation and AI will continue to drastically alter the work environment. This dynamic landscape confirms that the future of work is not static. The declining demand for formal degrees may impact the higher education sector, necessitating a rethink of educational models. Regardless of technological advancement, the 'human-in-the-loop' will remain essential to ensure ethical and effective work practices.

The future of work will transform the meaning and nature of work. Work will not be limited by geography and location. Hybrid, AI-augmented and remote work will increase accessibility to global talent and promote diversity across organizations.

The future of work will involve a heightened integration of technology and human intelligence. The future of work is changing as you read this book, and will continue to evolve based on social, geopolitical, economic, environmental, technological and the development of values and ethics. The new workspace will demand individuals to develop skillsets like self-discipline, life-long learning, commitment, and integrity, and not only technical know-how. The leader's role in this workplace will also be transformed from supervision and control to trust and guidance. The future of work belongs to those individuals and leaders who are able to collaborate, perceive technology as an enabler and grow sustainably.

References

Alessandrini, A., Campagna, A., Delle Site, P., Filippi, F., & Persia, L. (2015). Automated vehicles and the rethinking of mobility and cities. *Transportation Research Procedia, 5*, 145–160.

All about industries. (2024, September 25). Industrial robots in Germany 2023 was a record year for German automation companies. https://www.all-about-industries.com/german-industry-new-record-industrial-robots-2023-a-f2d99a46fd14e834100b17abe98ac0ef/?print

Amazon Canada. (2022). *The future of work: Employee expectations.*

Amazon. (2024). *Amazon's return to office policy.* https://www.aboutamazon.com/

Amery, B., & Bonen, T. (2018). *Future of work.* Labour Market Information Council. https://lmic-cimt.ca/wp-content/uploads/2020/01/LMI-Insights-Issue-2-EN.pdf

Arrieta, E. (2022). *The impact of COVID-19 on low-income households in the UK.* Institute for Fiscal Studies.

Attride-Stirling, J. (2001). Thematic networks: An analytic tool for qualitative research. *Qualitative Research, 1*(3), 385–405. https://doi.org/10.1177/146879410100100307

Automation Magazine. (2023). *Italy's robotics sector leads Europe.*

Bambra, C., Riordan, R., Ford, J., & Matthews, F. (2020). The COVID-19 pandemic and health inequalities. *Journal of Epidemiology and Community Health, 74*(11), 964–968. https://doi.org/10.1136/jech-2020-214764

Bank of Italy. (2022). *Annual report on the Italian economy.*

Barbier, E. B. (2016). Building the green economy. *Canadian Public Policy, 42*(S1), S1–S9. https://doi.org/10.3138/cpp.2015-017

Baryshnikova, N., Hall, S., & Wlaschiha, M. (2021). *Brexit and the pandemic: The double shock to the UK economy.* National Institute of Economic and Social Research.

Belgian Federal Government. (2023). *Right to disconnect and four-day working week initiative.* https://www.belgium.be/

Bielefeld, L., Fuchs, O., Pfeiffer, H., Lorenz, E., & Poslawskic, K. (2025). The importance of flexibility at work for Generation Z in Germany. *Journal of Economics, Management and Trade, 31*(9), 102-119. https://doi.org/10.9734/jemt/2025/v31i91352

Bloom, N., Barrero, J. M., Davis, S., Meyer, B., & Mihaylov, E. (2023, January 5). Research: Where managers and employees disagree about remote work. *Harvard Business Review.* https://hbr.org/2023/01/research-where-managers-and-employees-disagree-about-remote-work

Bloom, N., Han, R., & Liang, J. (2023). *How hybrid working from home works out.* National Bureau of Economic Research. https://doi.org/10.3386/w30292

Bloom, N., Han, R., & Liang, J. (2024). Hybrid working from home improves retention without damaging performance. *Nature, 622*, 49–55. https://doi.org/10.1038/s41586-023-06825-x

Blumer, H. (1969). *Symbolic interactionism: Perspective and method.* Prentice-Hall.

Boundless HQ. (2023). *Remote work in Germany: A guide for employers.* https://boundlesshq.com/guides/germany/remote-work/

Braun, V., & Clarke, V. (2006). Using thematic analysis in psychology. *Qualitative Research in Psychology, 3*(2), 77–101. https://doi.org/10.1191/1478088706qp063oa

Braverman, H. (1974). *Labor and monopoly capital: The degradation of work in the twentieth century.* Monthly Review Press.

Bridging tradition and flexibility. (n.d.). *Human Resource Excellent Magazine.* https://www.hr.com/en/magazines/all_articles/beyond-remote-work-ai-gig-economy-and-global-trend_mf14p3up.html

Britannica. (n.d.). *France.* https://www.britannica.com/place/France

Brookings Institution. (2021). *How COVID-19 is disrupting manufacturing supply chains.* https://www.brookings.edu/research/how-covid-19-is-disrupting-manufacturing-supply-chains/

Cabinet Erdyn, Cabinet Katalyse, DGEFP, Conseil Régional Hauts-de-France, Medef Lille Métropole, & Opcalia. (2019). *Impact of AI on jobs and skills in Hauts-de-France: Prospective study.* http://hauts-de-france.direccte.gouv.fr/sites/hauts-defrance.direccte.gouv.fr/IMG/pdf/synthese-etude-iahdf.pdf

Cameron, E. (2022). *Intelligent digital and the future of work.* Deloitte Insights.

Canada Beyond 150. (n.d.). *Future of work.* http://canadabeyond150.ca/assets/reports/Future%20of%20Work%20-%20EN.pdf

Carbonaro, G. (2022). *The 4-day workweek: A guide to the benefits and drawbacks.* Euronews.

Castleberry, A., & Nolen, A. (2018). Thematic analysis of qualitative research data: Is it as easy as it sounds? *Currents in Pharmacy Teaching and Learning, 10*(6), 807–815. https://doi.org/10.1016/j.cptl.2018.03.019

Catalan Engineers Worldwide. (n.d.) Smart Working in Italy. https://cew.cat/blog/smart-working-italy

Cath, C. (2018). Governing artificial intelligence: ethical, legal and technical opportunities and challenges. Philosophical Transactions of the Royal Society A: Mathematical, Physical and Engineering Sciences, 376(2133). https://doi.org/10.1098/rsta.2018.0080

Chan, C. (2021, September 25). Working out a global rethink: COVID-influenced work from home phenomenon likely to stick. *Vancouver Sun.* https://vancouversun.com/business/local-business/work-from-home-phenomenon-likely-to-stick

Charlton, E. (2024, November 22). *What is the gig economy and what's the deal for gig workers?* World Economic Forum. https://www.weforum.org/stories/2024/11/what-gig-economy-workers/

Chartered Institute of Personnel and Development. (n.d.). *AI algorithms, models, and systems.*

Christian, A. (2023, March 2). *The rise of the Gen Z side hustle.* BBC. https://www.bbc.co.uk/worklife/article/20230302-the-rise-of-the-gen-z-side-hustle

Clark Ashar, L. and Wallace, E. (2025, January 9). *Beyond remote work: AI, gig economy, and global trends shaping tomorrow's workplace.* https://www.hr.com/en/magazines/all_articles/beyond-remote-work-ai-gig-economy-and-global-trend_mf14p3up.html

Conference Board of Canada. (2020b, October). *Digital skills for today and tomorrow: Perspectives from Canadian employers and industry leaders.*

Conference Board of Canada. (2022, August 3). *Beyond blue and white collar: A skills-based approach to Canadian job groupings.* https://www.ecala.org/wp-content/uploads/2022/08/Conference-Board-of-Canada_beyond-the-blue-and-white-a-Skills-based-Approach-to-Can-Job-Groupings-Aug-22.pdf

Conference Board of Canada. (2022a, July 28). *From low-mobility to rapid-growth jobs: How governments and agencies can build the bridge to clean economy careers.*

Conference Board of Canada. (2022b, October). *The labour market of tomorrow: Projections from the model of occupations, skills, and technology (MOST).*

Conseil d'orientation pour l'emploi. (2017). *Automation, digitisation and employment report, Vol. 1: The impacts on employment volume, structure and location.* http://www.coe.gouv.fr/DetailNouveaute.html%3Fid_article=1347.htm

Coydel, F. (2022). *Short-term profit implications of digital transformation.* McKinsey & Company.

Cushman & Wakefield. (2021). *Future of workplace: Hybrid work models and office design.* https://www.cushmanwakefield.com/

Dabla-Norris, E., & Kochhar, K. (2018). *Gender, technology, and the future of work.* International Monetary Fund.

Davies, M. (2022). *IT infrastructure challenges in German companies.* Statista.

Davies, M. (2023). *Industry 4.0 and the future of manufacturing employment in Germany*. Statista.

De Lyon, J. & Dhingra, S. (2021, May 6). The impacts of Covid-19 and Brexit on the UK economy: early evidence in 2021. Issue 21 of Covid-19 analysis series. *Centre for Economic Performance*. London School of Economics & Political Science. https://cep.lse.ac.uk/_NEW/publications/abstract.asp?index=7982

Deloitte. (2021). *The future of work: Deloitte UK's hybrid working policy*. https://www2.deloitte.com/uk/en/pages/press-releases/articles/deloitte-announces-hybrid-working-model.html

Dengler, K., & Matthes, B. (2018). The impacts of digital transformation on the labour market: Substitution potentials of occupations in Germany. *Technological Forecasting and Social Change, 137*, 304–316.

Di Tecco, C., Ronchetti, M., Russo, S. Ghelli, M., Rondinone, B.M., Persechino, B., Iavicoli, S. (2021). Implementing Smart Working in Public Administration: a follow up study, National Library of Medicine, 112(2):141–152. doi: 10.23749/mdl.v112i2.10595, https://pmc.ncbi.nlm.nih.gov/articles/PMC8095324/

Dropbox. (2024). *Flexible work and organisational culture: The Dropbox way*. https://www.dropbox.com/

Easton E. and Djumalieva. J (2019) Creativity and the future of skills. London: Creative Industries Policy and Evidence Centre and Nesta. Available from: https://www.pec.ac.uk/research-reports/creativity-and-the-future-of-skills

Eckhardt, G. M., Houston, M. B., Jiang, B., Lamberton, C., Rindfleisch, A., & Zervas, G. (2019). Marketing in the sharing economy. Journal of Marketing, 83(5), 5-27. https://doi.org/10.1177/0022242919861929

Ellerbeck, S. (2022, June 24). *The Great Resignation is not over: Here's what the data says*. World Economic Forum. https://www.weforum.org/stories/2022/06/the-great-resignation-is-not-over/

Ellingrud, K., Gupta, R., & Salguero, J. (2020). *Building the vital skills for the future of work in operations.* McKinsey & Company.

Erdsiek, D., & Brautigan, S. (2022). *Hybrid work in the information economy: ZEW business survey.* ZEW - Leibniz Centre for European Economic Research.

Escape.com.au. (n.d.). *Mount Fuji and Tokyo Tower.*

Eures Europa. (2022). *Living and working in Italy.* European Commission.

Eures. (2023). *Labour market information: Italy.* European Commission.

European Commission. (2022). *France's recovery and resilience plan.* https://ec.europa.eu/info/business-economy-euro/recovery-coronavirus/recovery-and-resilience-facility/frances-recovery-and-resilience-plan_en

European Commission. (2023). *Economic forecast for Italy.*

European Commission. (2024, November 28). *France in the Digital Economy and Society Index.* https://digital-strategy.ec.europa.eu/

Excelsior Unioncamere/ANPAL. (2023). *Employment forecasts 2023–2027.*

Faber, E. (2025a, June 2). *Gen Zs and millennials at work: Pursuing a balance of money, meaning, and well-being.* Deloitte Insights. https://www.deloitte.com/us/en/insights/topics/talent/2025-gen-z-millennial-survey.html

Federal Statistical Office. (2023). *Population and demography.* https://www.destatis.de/EN/Themes/Society-Environment/Population/_node.html

Fields, A., Uppal, S., & LaRochelle-Côté, S. (2017). *The impact of aging on labour market participation rates.* Statistics Canada. https://www150.statcan.gc.ca/n1/en/pub/75-006-x/2017001/article/14826-eng.pdf

Filippucci, F., et al. (2025). *Macroeconomic productivity gains from artificial intelligence in G7 economies* (OECD Artificial Intelligence Papers No. 41). OECD Publishing. https://doi.org/10.1787/a5319ab5-en

Financial Times. (2022). *UK unemployment forecast to remain low despite economic headwinds.* https://www.ft.com/

Food and Agriculture Organisation of the United Nations. (2021). *COVID-19 and its impact on food and agriculture.* http://www.fao.org/2019-ncov/q-and-a/impact-on-food-and-agriculture/en/

Forbes. (2025, June 16). *How the creator economy is reshaping modern marketing – And why brands are paying attention.* https://www.forbes.com/sites/lowes-creator/2025/06/16/how-the-creator-economy-is-reshaping-modern-marketing--and-why-brands-are-paying-attention/

Fortuna, F., Rossi, L., Elmo, G. C., & Arcese, G. (2023) Italians and smart working: A technical study on the effects of smart working on the society, Technological Forecasting and Social Change, 187, 122220, ISSN 0040-1625, https://doi.org/10.1016/j.techfore.2022.122220

Franklin, N. (2021, August 6). *Managers and workers have different perception of the future 'employee experience'.* Workplace Insight. https://workplaceinsight.net/managers-and-workers-have-different-perception-of-the-future-employee-experience/

Frias, L. (2023). *California lawmaker reintroduces bill to make the 32-hour workweek the standard.* Business Insider.

Funke, C. (2023). *The global success of German discount retailers.* Handelsblatt.

Future Skills Centre (2022, August 3). *Beyond Blue and White Collar: a skills based approach to Canadian job groupings.* https://fsc-ccf.ca/research/beyond-blue-and-white-collar-a-skills-based-approach-to-canadian-job-groupings/

Garritsen De Vries, M. (1986, June 15). *The IMF in a Changing World.* International Monetary Fund. https://doi.org/10.5089/9780939934652.07

Gartner. (2023). *Human-machine collaboration in the future workplace.* Gartner Research. https://www.gartner.com/en/articles/human-ai-collaboration

Gartner. (n.d.). *Artificial intelligence terminology and definitions.* https://www.gartner.com/en/information-technology/glossary/artificial-general-intelligence-agi

General Electric. (2024). *Flexible work arrangements at GE.* https://www.ge.com/careers/

Gilbert-Johns, S., Mason, K., Ward, J., & Humby, L. (2022). *Migration trends in the UK: The combined effect of Brexit and COVID-19.* Migration Observatory.

Gilhooly, R. (2012, September 25). *Future of Japan's jobs market.* BCCJ Acumen. https://bccjacumen.com/future-of-japans-jobs-market/

Golden, L., Henly, J. R., Johnson, R. C., & Kleykamp, M. (2020). The elusive nature and measurement of remote work. *Journal of Business and Psychology, 35*(1), 1–14.

Goldman Sachs. (2023, April 19). *Technology: The creator economy could approach half-a-trillion dollars by 2027.* https://www.goldmansachs.com/insights/articles/the-creator-economy-could-approach-half-a-trillion-dollars-by-2027

Goldman Sachs. (2025). *Return to office policy announcement.* https://www.goldmansachs.com/

Google. (2024). *Google's work from anywhere policy and attendance requirements.* https://careers.google.com/

Gordon, J-S. (2021). Editorial: AI and law: ethical, legal, and socio-political implications. *AI & Society*, 36, 403–404. https://doi.org/10.1007/s00146-021-01194-0

Government Office for Science. (2017). *Future of skills and lifelong learning.* UK Government.

Gratton, L. (2022). *Redesigning work: How to transform your organisation and make hybrid work for everyone.* Pearson Education.

Gregory, R. W., & Valenti, A. (2021a). The rise of digital nomadism: Future directions and theoretical implications. *Journal of Business and Psychology, 36*(1), 1–14.

Gregory, R. W., & Valenti, A. (2021b). *The future of work: A review and research agenda.*

Gross, A., & Nolsoe, E. (2021, November 24). Back to work: Why French workers are resisting the Covid "Big Quit". *Financial Times.* https://www.ft.com/content/b7e77562-e7fb-4c29-b631-ad67bbe66a88

Gtmhub. (2023). *Agility in the modern enterprise.* https://gtmhub.com/

Guliyev, H., Huseynov, N., and Nuriyev, N. (2023) The relationship between artificial intelligence, big data, and unemployment in G7 countries: New insights from dynamic panel data model, World Development Sustainability, 3, 100107, ISSN 2772-655X, https://doi.org/10.1016/j.wds.2023.100107

Harari, D., Keep, M., & Brien, P. (2021). *Coronavirus: Economic impact.* House of Commons Library.

Harvard Business School. (2021). *Managing hybrid work: Strategies for success.* https://www.hbs.edu/

Hill, A. (2022). *The post-pandemic workforce: Trends and challenges.* The Guardian.

Hird, A. (2020, December 16). *Why France is struggling to embrace the work-from-home revolution.* RFI.

Hobbs, A. (2021, April 29). *The impact of remote and flexible working arrangements.* https://post.parliament.uk/the-impact-of-remote-and-flexible-working-arrangements/

Hochschild, A. R. (1983). *The managed heart: Commercialization of human feeling.* University of California Press.

Hofstede, G. (1980). Culture's Consequences: International Differences in Work-Related Values. Beverly Hills, CA: Sage Publications.

Horii, M., & Sakurai, Y. (2020). *The future of work in Japan: Accelerating automation after COVID-19.* McKinsey & Company.

HubSpot. (2023). *The 3E strategy for hybrid workplace success: Enablement, experimentation, empathy.* https://www.hubspot.com/

Hyder, S. (2021). *The future of work is employer-driven.* Inc. Magazine.

Iceland Technology Authority. (2021). *Results of the Icelandic four-day workweek trial 2015–2019*. https://www.government.is/

Institute for the Future of Work. (2023). *UK four-day workweek trials: Results and insights*. https://www.iFoW.org/

International Labour Organisation. (2019). *Skills mismatch in the labour market*. https://www.ilo.org/global/topics/skills/workforce-skills/skills-mismatch/lang--en/index.htm

International Labour Organisation. (2020). *Impact of COVID-19 on the world of work: Third edition*. https://www.ilo.org/global/topics/coronavirus/impacts/WCMS_743146/lang--en/index.htm

International Labour Organisation. (2021). *World of work report 2021: The role of digital labour platforms in transforming the future of work*. https://www.ilo.org/

International Monetary Fund. (2018). *Gender, technology, and the future of work* (IMF Staff Discussion Note).

Jaizuluddin, M., Pudji, H., Muhammad, F. R., Lambs, P. P., Irawan, S. S., Manifas, Z., Rizki, A. P., Andika, D. W., Yosa, P. S., & Angy, S. (2025). Technology foresight in Indonesia: Developing scenarios to determine electrical vehicle research priority for future innovation. *The Journal of Futures Studies, Strategic Thinking and Policy*, *27*(1), 120-132. https://doi.org/10.1108/FS-09-2023-0194

JLL Global Research. (2024, September 8). *The future of work survey 2024*. https://www.jll.com/en-uk/insights/future-of-work-survey

Johanson, M. (2022, May 11). *In places including the US and UK, remote work is here to stay. But that's not necessarily the story around the world.* BBC. https://www.bbc.com/worklife/article/20220511-the-countries-resisting-remote-work

Kalleberg, A. L., Reskin, B. F., & Hudson, K. (2000). Bad jobs in America: Standard and nonstandard employment relations and job quality in the United States. *American Sociological Review*, *65*(2), 256–278.

Kantchev, G. (2021). *Germany's labour shortage: The economic impact of an ageing workforce.* The Wall Street Journal.

Kazim, A. (2021). *Work-life balance in Southern Europe.* Journal of European Social Policy.

Kellogg Foundation. (2022). *Rethinking working time: Four-day workweek evidence and implications.* https://www.kelloggfoundation.org/

Kenny, R., Lunj, A., and Kis, A. (2025, 17 September). AI's thirst for water. Sustainable ICT. https://sustainableict.blog.gov.uk/2025/09/17/ais-thirst-for-water/

Kiger, M. E., & Varpio, L. (2020). Thematic analysis of qualitative data: AMEE Guide No. 131. *Medical Teacher, 42*(8), 846–854. https://doi.org/10.1080/0142159X.2020.1755030

Klasera, A., Cuel, R, & Casa, P. (2023). The future of hybrid work in Italy: A survey-based Socio-Technical-System analysis, *Journal of Innovation & Knowledge*, 8 (4). 100426. https://doi.org/10.1016/j.jik.2023.100426

Klaudijo Klaser, K., Cuel, R., & Casari, P. (2023). The future of hybrid work in Italy: A survey-based Socio-Technical-System analysis, Journal of Innovation & Knowledge, Volume 8, Issue 4, 2023, 100426, ISSN 2444-569X, https://doi.org/10.1016/j.jik.2023.100426

Kolade, O., & Owoseni, A. (2022). Employment 5.0: The work of the future and the future of work. *Technology in Society*, *71*, 102086. https://doi.org/10.1016/j.techsoc.2022.102086

Kolluru, M., Krishnan, K., & Kolluru, S. K. (2021). Post COVID-19 work strategies and implications: Insight on Indian IT sector. *Economics – Innovative and Economic Research Journal,* 9(2), 49-72.

Kolluru, M., Krishnan, S., & Kolluru, S. (2021). Strategic responses to the pandemic in the UK service sector. *Journal of Business Strategy*, *42*(3), 15–24.

Kondo, A. (2019). Impact of increased long-term care insurance payments on employment and wages in formal long-term care. *Journal of the Japanese and International Economies*, *53*, 101034. https://doi.org/10.1016/j.jjie.2019.101034

Korn Ferry. (2023). *Workforce 2024: Global insights report.* Korn Ferry Institute.

Kramer, K., & Wahl, H. W. (2015). *Demographic change in Germany.* Springer.

Kritikos, A. S., Mattes, A., Handrich, L., & Neumann, F. (2018). *The socio-ecological market economy: Germany's economic model.* Deutschland.de. https://www.reformgestaltung.de/fileadmin/user_upload/Dokumente/GIZ_Standard_A4_hoch_en_DIW_web.pdf

Kuchler, H. (2023). *CMR Surgical and the rise of robotic surgery.* Financial Times.

Kurian, J. S., & Bindu, M. N. (2024). Navigating the gig economy: Exploring challenges and motivations for the wellbeing of Gen Y and Gen Z gig workers. *Cogent Psychology, 11*(1). https://doi.org/10.1080/23311908.2024.2357458

Lamb, C., & Doyle, S. (2017). *Future-proof: Preparing young Canadians for the future of work.* Brookfield Institute for Innovation + Entrepreneurship.

Lancet Commission on the Future of Work. (2022). Advancing health and wellbeing in work environments. *The Lancet, 398*(10324), 398–405. https://doi.org/10.1016/S0140-6736(21)00096-6

Lazarova, M., Caligiuri, P., Collings, D. G., & De Cieri, H. (2023). Global work in a rapidly changing world: Implications for MNEs and individuals. *Journal of World Business, 58*(1), 101365. https://doi.org/10.1016/j.jwb.2022.101365

Lillico, A. (2017). *Federal actions for a clean growth economy delivering on the Pan-Canadian framework on clean growth and climate change.* Government of Canada. https://www.canada.ca/content/dam/themes/environment/documents/weather1/20170119-en.pdf

Liu, H., Li, N., Zhao, S., Xue, P., Zhu, C., and He, Y. (2025) The impact of supply chain and digitization on the development of environmental technologies: Unveiling the role of inflation and consumption in G7 nations, Energy Economics, 142, ISSN 0140-9883, https://doi.org/10.1016/j.eneco.2024.108165

LMiC. (2022, August 3). Future of Work, Beyond blue and white collar: A skills-based approach to Canadian job groupings. The Conference Board of Canada and the Future Skills Centre. https://lmic-cimt.ca/future-of-work/conference-board-of-canada-2022-august-3-beyond-blue-and-white-collar-a-skills-based-approach-to-canadian-job-groupings-the-conference-board-of-canada-and-the-future-skills-centre-2/

Loveridge, D. (2002). *The SEEPV acronym and process – A clarification.* Manchester, UK.

Loveridge, D., & Saritas, O. (2009). Reducing the democratic deficit in institutional foresight programmes: A case for critical systems thinking in nanotechnology. *Technological Forecasting and Social Change, 76*(9), 1208-1221. https://doi.org/10.1016/j.techfore.2009.07.013

Lund, S., Manyika, J., & Dua, A. (2019). *The future of work in America: People and places, today and tomorrow.* McKinsey Global Institute.

Mai, R. (2008). *The decline of the German family.* Population Studies.

Malhotra, A. (2021, March 23). *How "Future of Work" is helping solve Japan's workforce issues?* JungleWorks. https://jungleworks.com/how-future-of-work-is-helping-solve-japans-workforce-issues

Manyika, J., Chui, M., Miremadi, M., Bughin, J., George, K., Willicombe, P., & Dewhurst, M. (2017). *A future that works: Automation, employment, and productivity.* McKinsey Global Institute. https://www.mckinsey.com/

Mariappanadar, S. (2025, March 30). *Some Gen Zs are taking a 'micro-retirement'. It's one way to address burnout – but it comes with risks.* The Conversation. https://theconversation.com/some-gen-zs-are-taking-a-micro-retirement-its-one-way-to-address-burnout-but-it-comes-with-risks-252505

Market Research. (2025, October 4). Germany Robotics in Manufacturing Market. Ken Research. https://www.marketresearch.com/Ken-Research-v3771/Germany-Robotics-Manufacturing-44321214/

Martinetti, I. (2021, June 16). *Remote working is settling in for the long term in France, survey shows.* RFI.

Marx, K. (1867). *Das Kapital: Kritik der politischen Ökonomie [Capital: A critique of political economy]* (3 vols.). Otto Meissner. (Original work published 1867).

Maurer, R. (2025). *White House issues comprehensive workforce development strategy.* SHRM. https://www.shrm.org/topics-tools/news/trump-workforce-development-strategy

Mayer, J. D., & Salovey, P. (1997). What is emotional intelligence? In P. Salovey & D. J. Sluyter (Eds.), *Emotional development and emotional intelligence: Educational implications* (pp. 3–31). Basic Books.

McKinsey & Company (2025). *Flexible work's enduring appeal affects workers, employers, and real estate.* https://www.mckinsey.com/industries/real-estate/our-insights/flexible-works-enduring-appeal-affects-workers-employers-and-real-estate

McKinsey & Company. (2020a). *Digitalisation in Germany: Opportunities and challenges.* https://www.mckinsey.com/~/media/mckinsey/industries/healthcare%20systems%20and%20services/our%20insights/digitizing%20healthcare%20opportunities%20for%20germany/digitizing-healthcare-opportunities-for-germany.pdf

McKinsey & Company. (2020b). *Diversity wins: How inclusion matters.* https://www.mckinsey.com/featured-insights/diversity-and-inclusion/

McKinsey & Company. (2022). *The Great Attrition is making hiring even harder.* McKinsey Quarterly. https://www.mckinsey.com/

McKinsey & Company. (2023). *The future of work in Europe: Automation, workforce transitions, and the shifting geography of employment.* https://www.mckinsey.com/~/media/mckinsey/featured%20insights/future%20of%20organizations/the%20future%20of%20work%20in%20europe/mgi-the-future-of-work-in-europe-discussion-paper.pdf

McKinsey Global Institute. (2019a). *Notes from the AI frontier: Tackling Europe's gap in digital and AI.* https://www.mckinsey.com/~/media/McKinsey/Featured%20Insights/Artificial%20Intelligence/Tackling%20Europes%20gap%20in%20digital%20and%20AI/MGI-Tackling-Europes-gap-in-digital-and-AI-Feb-2019-vF.ashx

McKinsey Global Institute. (2019b). The future of work in America. McKinsey & Company. https://www.mckinsey.com/~/media/mckinsey/featured%20insights/future%20of%20organizations/the%20future%20of%20work%20in%20america%20people%20and%20places%20today%20and%20tomorrow/mgi-the-future-of-work-in-america-exec-summary-july-2019.pdf

Mercer. (2022). *Global talent trends: Italy.*

Mesaglio, M. (2025, November 24). *The New Rules of Human-AI Collaboration and Teamwork. Gartner.* https://www.gartner.com/en/articles/human-ai-collaboration

Miami, A. (2023). *The shrinking middle class and the future of work.* The Atlantic.

Microsoft Japan. (2020). *Work-life choice challenge: Summer 2019 pilot results.* https://news.microsoft.com/ja-jp/

Microsoft. (2021). *2021 work trend index annual report: The next great disruption is hybrid work. Are we ready?* https://www.brianheger.com/2021-work-trend-index-annual-report-the-next-great-disruption-is-hybrid-work-are-we-ready-microsoft/

Microsoft. (2022). *2022 work trend index: Annual report.* https://ms-worklab.azureedge.net/files/reports/2022/pdf/2022_Work_Trend_Index_Annual_Report.pdf

Microsoft. (2025). *Microsoft hybrid workplace policy and 2026 office return requirements.* https://www.microsoft.com/careers/

Ministry of Economy, Trade and Industry. (n.d.). *Digital policy.*

MIT Initiative on the Digital Economy. (2017). *The work of the future: Building better jobs in an age of intelligent machines.*

Moore, M., & Wilson, T. (2025, December 11). *Inside the failed green revolutions at BP and Shell.* Financial Times. https://www.ft.com/content/c6d7d225-f322-4e88-b7e3-2ca2beecd495

Moore, M., & Wilson, T. (2025, December 27). *How the energy giants tried to transform their businesses but ended up dramatically scaling back those plans and writing off billions of dollars*. Financial Times. https://financialpost.com/financial-times/inside-failed-green-revolutions-bp-shell

Muratovski, G. (2026, March 12). The Future Of Work Is Changing The Future Of Mobility, Forbes Technology Council. https://www.forbes.com/councils/forbestechcouncil/2026/03/12/the-future-of-work-is-changing-the-future-of-mobility/

National Post. (n.d.). *Toronto skyline.*

National Retail Federation. (2021). *Retail sales in 2021 expected to grow between 6.5 percent and 8.2 percent.* https://nrf.com/media-center/press-releases/

Newswire Canada. (2022, March 10). *OOO until TBD? Majority of Canadian office workers want remote work to stay*. https://www.newswire.ca/news-releases/ooo-until-tbd-majority-of-canadian-office-workers-want-remote-work-to-stay-897250807.html

Ng, M. A., Naranjo, A., Schlotzhauer, A. E., Shoss, M. K., Kartvelishvili, N., Bartek, M., ... & Silva, C. (2021). Has the COVID-19 pandemic accelerated the future of work or changed its course? Implications for research and practice. *International Journal of Environmental Research and Public Health, 18*(19), 10199. https://doi.org/10.3390/ijerph181910199

OECD. (2017). *Getting skills right: France.* OECD Publishing.

OECD. (2019). *Artificial intelligence in society*. OECD Publishing. https://doi.org/10.1787/eedfee77-en

OECD. (2025a). *OECD employment outlook 2025: Can we get through the demographic crunch?* https://doi.org/10.1787/194a947b-en

OECD. (2025b). *Unemployment rates: France, Germany, and the UK*. Organisation for Economic Co-operation and Development. https://www.oecd.org/en/data/indicators/unemployment-rate.html

OECD. (n.d.). *Gender wage gap*. https://data.oecd.org/earnwage/gender-wage-gap.htm

Office for National Statistics. (2022). *Long-term international migration, provisional: Year ending December 2020*. https://www.ons.gov.uk/

Office for National Statistics. (2023). *National population projections: 2021-based interim.* https://www.ons.gov.uk/

Okta. (n.d.). *Hybrid work and sustainability: Reducing the carbon footprint.* https://www.okta.com/

Osei, C., Gutu, M. and Hagan, D. (2026). The consumer: sustainable consumption perspectives. In Mena, S., Inoue, C.Y.A., Amaeshi, K., and Jayaram, D. *Handbook of Research on Sustainability and Governance* (p. 370), Edward Elgar Publishing. https://doi.org/10.4337/9781035328048

Owens, J. (2023). *Robotics in German manufacturing.* Industry Week.

Pakusch, C., Stevens, G., Boden, A., & Bossen, C. (2021). Unintended effects of autonomous driving: A study on mobility preferences in the future. *Sustainability, 10*(7), 2404.

Paruzel, A., Danel, M., & Maier, G. W. (2022). Scrutinizing the relationship between hybrid work characteristics and employee well-being. *Frontiers in Psychology, 13.*

Portuguese Parliament. (2023). *Law protecting rights of remote workers and establishing right to disconnect.* https://www.parlamento.pt/

PricewaterhouseCoopers. (2025). *Global Jobs Barometer 2025: Artificial intelligence impact on employment, wages, skills and productivity.* https://www.pwc.com/gx/en/services/ai/ai-jobs-barometer.html

Puschmann, T., & Alt, R. (2016). Sharing economy. Business & Information Systems Engineering, 58(1), 93-99

PwC. (2018). *Will robots steal our jobs? The potential impact of automation on the UK and other major economies.* PricewaterhouseCoopers.

PwC. (2022). *The future of work: A journey to 2022.* PricewaterhouseCoopers.

PwC. (2024). *Global workforce hopes and fears survey 2024.* PricewaterhouseCoopers.

PwC. (2025a). *Global workforce hopes and fears survey 2025.* PricewaterhouseCoopers.

PwC. (2025b). *Canada insights: AI jobs barometer.* https://www.pwc.com/gx/en/issues/artificial-intelligence/job-barometer/aijb-2025-canada-analysis.pdf

PwC. (2025c). *France insights: AI jobs barometer.* https://www.pwc.com/gx/en/issues/artificial-intelligence/job-barometer/aijb-2025-france-analysis.pdf

PwC. (2025d). *Germany AI jobs barometer: 2025 insights.* PricewaterhouseCoopers. https://www.pwc.com/gx/en/issues/artificial-intelligence/job-barometer/aijb-2025-germany-analysis.pdf

PwC. (2025e). *Italy AI jobs barometer: 2025 insights.* PricewaterhouseCoopers. https://www.pwc.com/gx/en/issues/artificial-intelligence/job-barometer/aijb-2025-italy-analysis.pdf

PwC. (2025f). *UK AI jobs barometer: 2025 insights.* PricewaterhouseCoopers. https://www.pwc.com/gx/en/services/ai/ai-jobs-barometer.html

Qwick. (2022). *Implementing the four-day work week in hospitality.* Qwick Insights.

Rao Madhu Sudhana, A., Kolluru, M., & Hyams-Ssekasi, D. (2022). Automation and the future of work in the UK. *International Journal of Manpower, 43*(2), 345–360.

Rehner, M. C., & Braeckeler-Kogel, V. (2022). *Germany: No right to remote work abroad for German employees.* Ius Laboris. https://iuslaboris.com/insights/no-right-to-remote-work-abroad-for-german-employees/

Reiter, T. (2022). *The retirement wave: Germany's demographic crisis.* Spiegel International.

Return to Work Research. (2021). *Return to the workplace: Research on employee preferences and wellbeing.* https://returntoworkresearch.org/

Robards, J. (2024). *Population dynamics in the UK: Births, deaths, and the pandemic effect.* Centre for Population Change.

Robinson, B. (2025, January 29). *'Micro-Retirement': The New Career Trend Rising Among Gen Z.* Forbes.

https://www.forbes.com/sites/bryanrobinson/2025/01/29/micro-retirement-the-new-career-trend-rising-among-gen-z/

Rodrigues, A. (2022). *Working from anywhere: Legal implications for global employers.* Global HR Insights.

Rome Business School. (2022). *Women in the Italian workforce: Economic impact.*

Rose, N. (1999). *Powers of freedom: Reframing political thought.* Cambridge University Press; 1999.

Rosenblat, A., & Stark, L. (2021). Algorithmic management: A framework for future work and research. *Proceedings of the ACM on Human-Computer Interaction, 5*(CSCW1), 1–22.

Sadie, J. (2025 January 27) 2024 Future of Work Summit: How Amazon's upskilling program helps prepare employees for the future. https://www.benefitscanada.com/events/conference-coverage/2024-future-of-work-summit-how-amazons-upskilling-program-helps-prepare-employees-for-the-future/

Salesforce. (2024). *Salesforce's flexible, human-centric workplace approach.* https://www.salesforce.com/company/careers/

Schor, J. (2016). Debating the sharing economy. Journal of Self-Governance and Management Economics, 4(3), 7-22. https://www.tellus.org/pub/Schor_Debating_the_Sharing_Economy.pdf

Scottish Government. (2023). *Four-day working week trial: Results and future implementation.* https://www.gov.scot/

Seike, A. (2017, November 13). *The future of work under socioeconomic structural changes* [Interview]. International Labour Organisation. https://www.ilo.org/tokyo/FoW/WCMS_595667/lang--en/index.htm

Sennett, R. (1998). *The corrosion of character: The personal consequences of work in the new capitalism.* W.W. Norton & Company.

Shopify. (2024). *Remote-first company culture: Building connection across distributed teams.* https://www.shopify.com/

Shutterproof. (n.d.). *Leaning Tower of Pisa.*

Shuttleworth, A., Schmitz, S., & Beier, G. (2022). The impact of Industry 4.0 on the German manufacturing workforce. *International Journal of Production Economics.*

Shuttleworth, L., Schmitz, S., & Beier, G. (2022). Impacts of Industry 4.0 on industrial employment in Germany: A comparison of industrial workers' expectations and experiences from two surveys in 2014 and 2020. Production & Manufacturing Research, 10(1), 583–605. https://doi.org/10.1080/21693277.2022.2104400

Simon, E. (2025, January 29). *Earlier 'micro-retirement' could lead to bigger pension pots: Standard Life.* Corporate Adviser. https://corporate-adviser.com/earlier-micro-retirement-could-lead-to-bigger-pension-pots-standard-life/

Smith, A. (1776). *An inquiry into the nature and causes of the wealth of nations.* W. Strahan & T. Cadell.

Somers, M. (2022). *The skills gap is widening.* MIT Sloan Management Review. https://sloanreview.mit.edu/landing_page/the-future-of-workplace-learning/

Spanish Government. (2023). *32-hour working week pilot programme.* https://www.lamoncloa.gob.es/

Spotify. (2023). *Work from anywhere policy outcomes and diversity impact.* https://www.spotify.com/

Standing, G. (2011). *The precariat: The new dangerous class.* Bloomsbury Academic.

Statista Research Department. (2023). *Gender pay gap in OECD countries.*

Statista. (2022). *Japan: Total population 2017-2027.* https://www.statista.com/statistics/263746/total-population-in-japan/

Statista. (2023). *Population of Germany by age group.*

Statista. (2025). *Total population of France from 1982 to 2025.* https://www.statista.com/statistics/459939/population-france/

Statistics Canada. (2021). *Canadian social survey: Well-being, unpaid work and family life.* https://www.statcan.gc.ca/en/survey/household/5354-WUWFT

Subramaniam, V. (2022, September 22). Canadians, especially men, want remote work to stick around. *The Globe and Mail.* https://www.theglobeandmail.com/business/article-remote-work-preference-canada/

Sullivan, S. E., & Arthur, M. B. (2006). The evolution of the boundaryless career concept: Examining physical and psychological mobility. *Journal of Vocational Behavior, 69*(1), 19–29.

Suresh, B. L (2023). Demographic and Socio-Economic Development — Evidence from G7 Countries, Paradigm Academic Press Studies in Social Science & Humanities ISSN 2709-7862 AUG. 2023 VOL.2, NO.8, doi:10.56397/SSSH.2023.08.03

Tangalakis-Lippert, K. (2025, June 19). *Gen Z is bringing the gig economy to corporate America.* Business Insider. https://www.businessinsider.com/gen-z-gig-economy-mindset-corporate-america-2025-6

Taylor, F. W. (1911). *The principles of scientific management.* Harper & Brothers.

The Guardian. (2022). *The four-day week pilot: Results and implications.* https://www.theguardian.com/

The Week. (n.d.). *Out of office: Microretirement is trending in the workplace – Long vacations are the new way to beat burnout.* https://theweek.com/business/jobs/microretirement-workplace-trend-jobs-employment

Theodorou, A. (2025, December). 4 workforce management moves retailers need for 2026 (and beyond), https://www.retailgazette.co.uk/blog/2025/12/4-workforce-management-moves-retailers-need-for-2026-and-beyond/

Thompson, P. and Warhurst, C. (1998). *Critical Perspectives on Work and Organisations Series*, London, MacMillian Press Limited.

Thorogood, J. (2025, July 16). *Gig economy in 2025: Regulatory shifts and tech-driven opportunities*. Forbes. https://www.forbes.com/councils/forbesbusinesscouncil/2025/07/16/gig-economy-in-2025-regulatory-shifts-and-tech-driven-opportunities/

Tirado Robles, C. (2022). The report "Future of Work 2035: For everyone to shine": Legal analysis of the Japanese proposals on the implementation of robots and AI in the world of work and their possible adaptation to the European context. *Mirai: Estudios Japoneses*, *6*, 205–216.

TUC. (2020 November 25). Impact of Covid-19 and Brexit for the UK economy. Review of forecasts published between July-November 2020. https://www.tuc.org.uk/research-analysis/reports/impact-covid-19-and-brexit-uk-economy

UK Hospitality. (n.d.). The four-day working week: how does it work? https://www.ukhospitality.org.uk/the-four-day-working-week-how-does-it-work/

UNESCO. (2021). *COVID-19 impact on education: Managing disruption, responding to crisis*. https://www.unesco.org/

United Nations Development Programme. (2021). *Global impact of COVID-19: Economic and social consequences*. UNDP.

US Department of Labor. (2025). *America's talent strategy: Building the workforce for the golden age*. https://www.dol.gov/sites/dolgov/files/OPA/newsreleases/2025/08/Americas-Talent-Strategy-Building-the-Workforce-for-the-Golden-Age.pdf

Vadavi, S., & Sharmiladevi, J. C. (2024). Analysing the evolution and patterns of the gig economy: A bibliometric examination of growth and trends. *Cogent Business & Management*, *11*(1). https://doi.org/10.1080/23311975.2024.2424480

Vaismoradi, M., & Snelgrove, S. (2019). Theme in qualitative content analysis and thematic analysis. *Forum: Qualitative Social Research*, *20*(3), Article 23. https://doi.org/10.17169/fqs-20.3.3376

Van Eerd, R., & Guo, J. (2020, January 17). *Jobs will be very different in 10 years. Here's how to prepare*. World Economic Forum. https://www.weforum.org/stories/2020/01/future-of-work/

vom Brocke, J., Winter, R., Hevner, A., & Maedche, A. (2018). *Design science research cases*. Springer.

Walker, C. and Goldsmith, I. (2026, February 27). From Energy Use to Air Quality, the Many Ways Data Centers Affect US Communities. World Resources Institute. https://www.wri.org/insights/us-data-center-growth-impacts

Warhurst, C. and Thompson, P. (2006) Mapping knowledge in work: proxies or practices? Work, Employment & Society; 20; 787 http://wes.sagepub.com DOI: 10.1177/0950017006069815

Weber, M. (1905). *The Protestant ethic and the spirit of capitalism* (T. Parsons, Trans.). Charles Scribner's Sons.

Whitlock, R., Spencer, A., & Barr, S. (2024). *Cultural heritage and urban tourism in London*. Routledge.

Williams, A., & Warhurst, C. (2020b). In praise of care work: A response to the COVID-19 pandemic. *Work, Employment and Society*, *34*(1), 151–158.

Wolf, M. (2022). *The crisis of British capitalism*. Financial Times.

Wooldridge, A. (2023). *The end of the German economic miracle*. Bloomberg.

World Economic Forum. (2018). *The future of work: Employment, skills and workforce strategy for the fourth industrial revolution*. https://www.weforum.org/

World Economic Forum. (2021). *The future of jobs report 2021*. https://www.weforum.org/reports/the-future-of-jobs-report-2021/

World Economic Forum. (2023). *The future of jobs report 2023*. https://www.weforum.org/

World Economic Forum. (2025, January 7). *The Future of Jobs Report 2025*. https://www.weforum.org/publications/the-future-of-jobs-report-2025/in-full/3-skills-outlook/

World Economic Forum. (2025a). *Future of jobs report 2025: Generational perspectives on work transformation.* https://www.weforum.org/

World Economic Forum. (2025b). *The future of jobs report 2025.* https://www.weforum.org/publications/the-future-of-jobs-report-2025

World Health Organisation. (2021). *COVID-19 pandemic: Global epidemiological overview.* WHO.

World Population Review. (2023). *Italy population 2023.* https://worldpopulationreview.com/countries/italy-population

Worldometer. (2023a). *Canada population.* https://www.worldometers.info/world-population/canada-population/

Worldometer. (2023b). *Germany population.* https://www.worldometers.info/world-population/germany-population/

Worldometer. (2023c). *Italy demographics.* https://www.worldometers.info/demographics/italy-demographics/

Wright, M. (2022). *Retail management in the hybrid era.* Retail Gazette.

Wright, O. (2023). *THG orders staff back to the office.* The Times.

Wrzesniewski, A., & Dutton, J. E. (2001). Crafting a job: Revisioning employees as active crafters of their work. *Academy of Management Review, 26*, 179-201. https://doi.org/10.2307/259118

Wrzesniewski, A., & Dutton, J. E. (2021a). *Revisioning employees as active crafters of their work.* Academy of Management Review.

Wrzesniewski, A., & Dutton, J. E. (2021b). Crafting work design for meaningful work. *Annual Review of Organizational Psychology and Organizational Behavior, 8*(1), 21–42.

Xiao, T., Nerini, F.F., Matthews, H.D 2025. Environmental impact and net-zero pathways for sustainable artificial intelligence servers in the USA. Nat Sustain 8, 1541–1553. https://doi.org/10.1038/s41893-025-01681-y

Word Index

www.ingramcontent.com/pod-product-compliance
Lightning Source LLC
LaVergne TN
LVHW011814200926
848484LV00005B/1198